P9-DTD-615

FROM COPERNICUS TO EINSTEIN

HANS REICHENBACH

Translated by
RALPH B. WINN

Dover Publications, Inc.
New York

Copyright © 1942, 1970 by Philosophical Library, Inc.
All rights reserved under Pan American and International Copyright Conventions.

Published in Canada by General Publishing Company, Ltd., 30 Lesmill Road, Don Mills, Toronto, Ontario.
Published in the United Kingdom by Constable and Company, Ltd., 10 Orange Street, London WC2H 7EG.

This Dover edition, first published in 1980, is an unabridged and corrected republication of the English translation first published in 1942 by Philosophical Library, Inc.
The work was originally published in 1927 by Ullstein, Berlin, under the title *Von Kopernikus bis Einstein.*

International Standard Book Number: 0-486-23940-3
Library of Congress Catalog Card Number: 79-055911

Manufactured in the United States of America
Dover Publications, Inc.
180 Varick Street
New York, N.Y. 10014

CONTENTS

ILLUSTRATIONS

FROM COPERNICUS TO EINSTEIN

Chapter 1 : THE COPERNICAN VIEW OF THE WORLD

THIS little book purports to serve as an introduction to the great problems of space, time and motion. The inquiries it is concerned with are very old. Men have been forming ideas concerning space and time since times immemorial, and curiously enough, have been writing and fighting about these things with the greatest interest, even fanaticism. This has been a strange strife, indeed, having little to do with economic necessities; it has always dealt with abstract things, far removed from our daily life and with no direct influence upon our daily activities. Why do we need to know whether the sun revolves around the earth or vice versa? What business of ours is it, anyway? Can this knowledge be of any use to us?

No sooner have we uttered these questions than we become aware of their foolishness. It may not be of any use to us, but we want to know something about these problems. We do not want to go blindly through the world. We desire more than a mere existence. We need these cosmic perspectives in order to be able to experience a feeling for our place in the world. The ultimate questions as to the meaning of our actions and as to the meaning of life in general always tend to involve astronomical problems. Here lies the mystery surrounding astronomy, here lies the wonder we experience at the sight of the starry sky, the wonder growing in proportion to our understanding of immense distances of space and of the stars' inner nature. Here is the source of scientific as well as popular astronomy.

These two branches have diverged in the course of their development. Astronomy, as a science, has come to forget its primitive wonder: instead, it approaches the realm of stars with sober research and calculation. This disenchantment with its subject-matter, which scientific study invariably entails, has permeated astronomy to a greater degree than the layman realizes. In observing the astronomers of today, how they measure, take notes, calculate, how little attention they pay to mysterious speculations, one may be surprised to find the wonderful structure of learning so cut and dry at a close range. Yet nothing is more wrong and more objectionable than the feeling of a heartbreaking loss, with which some people regard the vanishing mysticism of the skies. Although science may have destroyed a few naive fantasies, what she has put in their place is so immensely greater that we can well bear the loss.

It takes perseverance and energy, of course, to comprehend the discoveries of science; but whoever undertakes the study is bound to learn many more surprising things from it than a naive study of nature can disclose. Scientific astronomy has always exercised, in fact, a great influence upon everyday thinking and upon the popular conception of the universe. If it is difficult today to pronounce the name of Copernicus without thinking of a turning point of history, it is not only because the name is connected with a profound transformation in the science, but also because all our knowledge and thinking have been deeply affected by his discovery. The statement that the earth does not occupy the center of the world means more than an astronomical fact; we interpret it as asserting that man is not the center of the world, that everything which appears large and mighty to us is in reality of the smallest significance, when measured by cosmic standards. The statement has been made possible as a result of scientific development in the course of thousands of years, yet it definitely contradicts our immediate experience. It takes a great deal of training in thinking to believe in it at all. Nowadays we are no longer conscious of these things, because we have been brought up since childhood in the Copernican view of the world. However, it cannot be denied that the view belies the testimony of our senses, that every immediate evidence shows the earth as standing still while the heavens are moving. And who among us can declare in all seriousness that he is able to imagine the tremendous size of the sun or to comprehend the cosmic distances defying all earthly ways of measurement? The significance of Copernicus lies precisely in the fact that he broke with an old belief apparently supported by all immediate sensory experiences. He could do it only because he had at his disposal a considerable amount of accumulated scientific thought and scientific data, only by Ptolemy Claudius of Alexandria and outlined in his famous work *Almagest.* The most important feature of the Ptolemaic scheme of the universe is the principle that the earth is the center of the world. The heavenly globe revolves around it; and Ptolemy knew full well that it has the same spherical shape below the horizon, which it assumes above the horizon. In fact, Ptolemy knew even that the earth is a sphere. His proofs to this effect reveal a great knowledge of astronomy. He shows, first of all, the existence of curvature from north to south. As the Polar Star stands higher in the north and lower in the south, the surface of the earth must be correspondingly curved. The proof of the existence of curvature from west to east reveals even better observation. When the clocks are set by the sun in two places located west and east, and when an eclipse of the moon is thus observed, it will be seen at different times. However, the eclipse is a single objective event and should be seen everywhere at the same time. Hence we conclude that the clocks at the two places are not in accord. This can be accounted for by the curvature of the earth in the west-east direction: the sun passes the line of the meridian at different moments in different places.

In spite of the recognition of the spherical shape of the earth, Ptolemy was far from admitting its movement. He contended, on the contrary, that it was impossible for the earth to be moving at all, either in a rotating or in a progressive manner. As far as the former is concerned, he admitted the possibility of such an opinion, as long as the movement of the stars was considered. However, when we take into consideration everything that happens around us and in the air, this view—so he argues—becomes obviously absurd. For the earth, during its rotation, would have to leave the air behind. Objects in the atmosphere, such as flying birds, not being able to follow the rotation, would have to be also left behind. A progressive motion of the earth is equally impossible for, in that case, the earth would leave the center of the heavenly sphere, and we would see by night a smaller part of the sphere and by day a larger one.

One can see from these arguments that the great astronomer has devoted much serious thought to the problem. In the light of his rather limited knowledge of mechanics and of the heavenly spaces, his reasoning must have seemed quite conclusive. As far as his last objection was concerned, he could not have suspected that the interstellar distances were so great as to make the lateral shift of the earth completely unnoticeable.

The planets are characterized, according to Ptolemy, by common movements. Their path, as observed in the sky, is determined by superimposed circular orbits. As a result, there arise the so-called "epicycles." One must admit that Ptolemy has deeply understood the nature of planetary movements. When one gets acquainted with the Copernican conception, one discovers the facts revealed behind Ptolemy's epicycles: the loop of the planets' course mirrors their double motion as regards the earth. In the first place, they move in a circle around the sun, and in the second place, this movement is observed from the earth which, in its turn, revolves around the sun.

The Ptolemaic conception of the universe dominated the learned people's minds for more than one thousand years. The man who undermined this firm tradition—Nicholas Copernicus—required great independence of thought as well as great scientific knowledge, for only an insight into the ultimate relations of nature could give him the ability to discern new approaches to truth.

The canon of Frauenburg was long known as a learned astronomer before his new ideas were presented; he had studied in Italy all branches of science, he had acted as doctor and church administrator in his home town, and his astronomic knowledge was so well recognized that in 1514 he was asked by the Lateran Council for his opinion on questions of calendar reform. His new ideas concerning the system of the universe were formed, in their essence, at the age of 33. However, he did not promulgate them at that time, but devoted the following years to a thorough elaboration and demonstration of his theories. Only excerpts of his doctrine were published during his lifetime. His main work entitled "Of the Rotation of Celestial Bodies" appeared only after his death in 1546. He read the proofs only on his death-bed and thus failed to notice that his friend Osiander supplied the work with a foreword which contained a cautious compromise with the opinions of the Church.

If we examine the proofs given by Copernicus of his

new theory, we find them quite insufficient from the point of view of present-day knowledge. He was able, in fact, to cite as a distinct advantage only the greater simplicity of his system. He regards it as improbable that the stars move with great speed in their large orbits and finds it more likely that the earth rotates on its axis, so that the speed of motion in each particular point is considerably smaller. Against Ptolemy's objection to this he urges that Ptolemy considered the rotating movement of the earth as implying force, whereas it is simply natural; its laws differ completely from those of a sudden jerky movement. All of this is certainly inconclusive. We know today that Newton's theory contains the first real proof of the Copernican conception of the universe. But it seems that new ideas are able to gain foothold by the sheer power of their inherent truth long before their objective verification has been obtained.

On the other hand, it is very important to acknowledge that the Copernican theory offers a very exact calculation of the apparent movements of the planets and that the tabulations (the so-called "Ephemerides") accompanying it are far superior to the older ones. Here lies one of the reasons which led the scientists to accept the Copernican system, even though it must be conceded that, from the modern standpoint, practically identical results could be obtained by means of a somewhat revised Ptolemaic system. Furthermore, Copernicus calculated quite accurately the radii of the planetary orbits (within less than 1%). In fact, he knew already that the sun must be slightly off the center of the solar system, for an assumption to the contrary led to estimable discrepancies.

Yet there was still a long way from this discovery to the recognition of the elliptic shape of the orbits; any conclusive evidence to this effect required above all better astronomic instruments. In this important connection, we must consider Tycho Brahe who is less prominent as a theoretician than as a builder of outstanding instruments. Brahe was able to work for many decades under the protection of the Danish king. He built the castle Uranienburg on an island, to which was attached a large settlement where precise instruments were prepared for him in special plants. It is amazing how the precision of instruments was increased in this manner. For instance, Copernicus had to be satisfied with measurements within 10' of the arc. This corresponds approximately to an angle covered by a five-pence piece at a distance of six meters. Tycho increased the precision to within half a minute of the arc. This angle would be enclosed by the same coin at a distance of 120 meters. With the instruments of today, of course, angles can be measured within one hundredth of a second of the arc. The coin would have to be placed at a distance of 360 kilometers to enclose such a small angle.

This precision we owe mainly to the use of the telescope. Tycho had to work without a telescope. One of his sextants with which he conducted his observations of Mars still stands in the Prague observatory, where Tycho, exiled from Denmark, spent the last years of his life (c. 1600).

Figure 1 shows the picture of this historic instrument. The pointed leg is set in a stand. The whole instrument is movable at the hinge in the upper end of the leg. It measures 1½ meter at the shank. The shank may be turned and has a sight-hole at the bottom to the left, an ironplate with a slit, through which a sharp edge on the

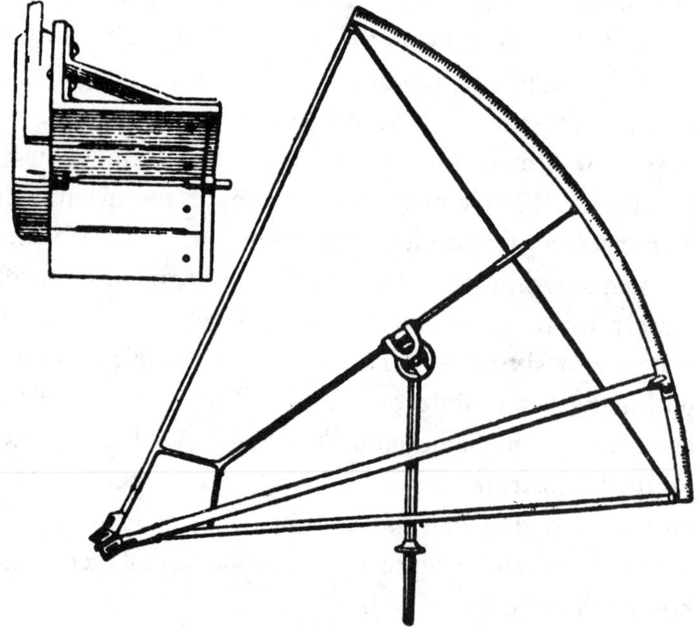

Figure 1. *A Tycho Brahe's Sextant*

upper end of the shank (to the right) is adjusted. This endpiece slides along an angle-scale. The sight-plate itself measuring several centimeters is reproduced in an enlarged form at the upper left corner. By means of such a crude-looking apparatus, Tycho found the data on which modern astronomy is historically resting.

The man who continued Tycho Brahe's work was his assistant Johann Kepler whose name surpasses by far that of his master. Kepler carried on his observations with the sextants of Tycho. He determined the course of the motion of Mars by means of so many individual observations that he was able to pronounce it with certainty as elliptical in shape. He discovered through mere measurement also other laws of planetary motion, called after him "the Kepler's laws." One must admire the strength of character of this man, which manifests itself in his zeal for factual accuracy. Kepler was at first a mystic and speculative dreamer, disinclined to sober observations. He concentrated in his early works on searching for strange mathematical 'harmonies' of nature, and such a goal inclines one to distort facts rather than to establish them. It remains true, however, that Kepler has accomplished much more for his own aim by his zeal for factual accuracy than by his speculations. He himself expresses this thought. In his work entitled "Harmony of the World," which appeared in 1619, he writes concerning the discovery of his laws: "At last I have found it, and my hopes and expectations are proven to be true that natural harmonies are present in the heavenly movements, both in their totality and in detail—though not in a manner which I previously imagined, but in another, more perfect, manner. . . If you forgive me, I shall be glad; if you are angry, I shall endure it. Here I cast my dice and write a book to be read by my contemporaries or by the future generations. It may wait long centuries for its reader. But even God himself had to wait for six thousand years for those who contemplate his work."

We must not forget, however, that, though the astronomic picture of the universe was considerably advanced, in regard to precision, by Kepler's discoveries, nevertheless, that world-view, though basically Copernican, differed very considerably from our Copernican idea of the world. Copernicus as well as Kepler was of the opinion that the solar system virtually exhausted the space of the universe. The stars, according to them, were tiny dots in the sphere of heavenly matter, which circumscribed the whole of space. When Giordano Bruno expressed his thoughts on the infinity of the firmament and maintained that fixed stars were independent solar systems, Kepler proceeded immediately to combat the idea. How difficult it must have been to climb the stairs leading to our present-day knowledge!

Astronomy made its decisive advance over Kepler's knowledge again through an improvement in the means of observation — through the invention of the telescope. The great merit of having made the first serviceable telescope and of having used it for the observation of the sky belongs to Galileo; though not the original inventor of the telescope, he constructed it after hearing of such instruments. He directed his telescope toward the moon and recognized the spots on the moon, on account of their jagged outline and shifting illumination, as tremendous mountains (1610). He pointed it towards Venus and saw its sickle-like shape, similar to that of the moon, which it periodically assumes as a result of receiving light from the sun. He directed the telescope towards Saturn and saw its 'triple' figure the details of which he could not yet discern. He directed it towards Jupiter and saw its satellites (the four brighter ones) designated by him as "medizeic planets."

All these facts, with their enlargement and enrichment of the Copernican world, must have greatly astonished his contemporaries. It also provoked, to be sure, the opposition of the old school of scientists who saw their tenets grounded in Aristotle seriously endangered. Galileo's most precarious position can be best envisaged from a letter written by him to Kepler: "I am very grateful that you have taken interest in my investigations from the very first glance at them and thus have become the first and almost the only person who gives full credence to my contentions; nothing else could be really expected from a man with your keenness and frankness. But what will you say to the noted philosophers of our University who, despite repeated invitations, still refuse to take a look either at the moon or the telescope and so close their eyes to the light of truth? This type of people regard philosophy as a book like Aeneid or Odyssey and believe that truth will be discovered, as they themselves assert, through the comparison of texts rather than through the study of the world or nature. You would laugh if you could hear some of our most respectable university philosophers trying to argue the new planets out of existence by mere logical arguments as if these were magical charms." Galileo relates how another scientist refused to take a look through the telescope "because it would only confuse him." The tragic fate of Galileo, caused by such antagonism, is well known. He had to pay with many years of incarceration and imprisonment for his sponsorship of the Copernican theory.

Another achievement of Galileo had apparently no direct connection with astronomy; but this connection was discerned soon enough. Galileo was the first man to investigate the laws of falling bodies. He has thereby established the basic laws on which the science of mechanics was destined to grow. The apparatus he built was quite primitive. For instance, he had no watch in the modern sense of the word, but had to measure time by means of water running out of a vessel. In spite of everything, he was able to determine the relationship between the distance and the time of the fall, and also the law of acceleration. He also discovered the fact — a most surprising fact for his day — that all bodies fall equally fast. Finally, he formulated the basic law of motion, named after him: that every body unaffected by external forces moves in a straight line at a uniform speed, and that this motion can never stop by itself.

Although these laws seem to be merely bits of factual information, nevertheless they signify an extraordinary progress as compared to the preceding era. There was no inclination at that time to collect data. It was believed that all one wanted to learn could be disclosed by speculative thinking. Galileo's great achievement was that he resorted to direct investigation of nature. Moreover, the facts he discovered were destined to attain a significance far beyond their own realm, namely, when Newton constructed the mechanics of heavens on them.

Fate allotted to the English physicist Isaac Newton (1643-1727) an outstanding role in the history of the natural sciences of the described period. He was the great unifier who combined the individual discoveries of Copernicus, Kepler and Galileo into one magnificent system. His intellectual achievement cannot be estimated too highly. With the vision of a genius he realized that the power of gravitation perceived by Galileo in his doctrines concerning falling bodies had a significance far transcending the region of the earth, that this power of attraction constituted a property of all mass, and that it determined the planets' behavior across cosmic distances. This far-reaching insight into the nature of things was accompanied by Newton's great caution in scientific investigation. He started with the correct premise that the power of attraction must diminish with distance. He then calculated what the magnitude of this power, already estimated by Galileo on the surface of the earth, could be at the distance of the moon. Next he computed the length of time required for the revolution of the moon around the earth, if this gravitational power was indeed responsible for the motion of the moon. All this was a magnificent elaboration of the original idea. Unfortunately, luck was against Newton, and his investigations resulted in anything but agreement with facts. Yet nothing shows better the greatness of the scholar's character than his conduct in the face of failure: he put his calculations away in a closet without publishing a single word concerning his profound meditations (1666). Only twenty years later could the mistake be explained. The length of the earth's radius, taken by Newton as the basis of his calculations, had been inexact; new estimates on the astronomers' part gave a new measurement with which Newton's reflections about the moon proved to be in full accord.

The mechanics of Newton has thus received confirmation, and it must have seemed like a magic key to his contemporaries. His theory transformed the fundamental facts of the preceding centuries into a uniform system, including the Copernican theory of the heliocentric motion of the planets, Kepler's laws concerning their orbits, and Galileo's laws of falling bodies in a gravitational field. Kepler did not live to greet this triumph of thought; no doubt, he would have rejoiced over this proof of the harmony of cosmic motions.

The Copernican conception of the universe was at last scientifically established, insofar as the laws underlying it stood revealed. Up to that time the Copernican conception of the universe, as compared to the Ptolemaic conception, could justify itself only by its claim of representing the world-picture in simpler terms. But now, with the addition of Newtonian mechanics, it became the only acceptable one. Its real merit was made explicit: the Copernican conception of the world provided an explanation of natural phenomena, a cosmic order governed by laws. It was the destiny of the Western mind to absorb this worldview which so much corresponded to its innate tendencies of thought.

Thus ends the first period of new physics; and with it has come a new method of inquiry to dominate the natural sciences ever since. The collection of facts is the starting point of investigation; but it does not mark its end. Only when an explanation comes like a bolt of lightning and melts separate ideas together in the fire of thoughtful synthesis, is that stage reached which we call understanding and which satisfies the seeking spirit.

The following chapters will show how widely and how consciously new physics has carried through this method of inquiry.

Chapter 2 : ETHER

WE HAVE already pointed out, in connection with the Copernican picture of the world, that the astronomic problems of motion and gravitation represent one of the sources from which the theory of relativity has sprung. Its other source lies in the theory of electricity and in that of light. We shall now concern ourselves with its development from this latter source; and in so doing, we shall follow the trend of development characterizing the modern conception of the physical universe. The truth is that the science of physics was forced to go beyond the views of Copernicus, Galileo and Newton by questions arising in connection with electricity and optical phenomena. These men, considered as innovators at their time, experienced all the inimical resistance of an outworn age still fighting for its existence, as we can judge from Galileo's tragic words quoted above. For the succeeding period the same men represent the classics, the great authorities who have dominated the thoughts of a whole era and whose work was carried on by generations of scholars; and the younger generation has to fight against them a battle similar to that which made those men famous.

It seems that progress in the knowledge of nature can be made only through conflict between two successive generations. What is considered at one time as a revolution of all thinking, a tempest in the brain, is for the next age a matter of fact, a school knowledge acquired under the influence of one's environment and believed and proclaimed with the certainty of everyday experience. Thus, possible criticism to which even the greatest discoveries should be continuously submitted, is forgotten; thus we lose sight of the limitations holding for the deepest insights; and thus man forgets in his absorbing concern with the particulars to re-examine the foundations of the whole structure of knowledge. We shall always have to depend on men like Copernicus who question obvious matters and whose critical judgment penetrates deep into the foundations of truth.

The history of the study of light illustrates this process. For it represented a definite attempt to comprehend the phenomena of light on the basis of ideas aroused by new astronomy and mechanics; it was an attempt to make mechanics the last court of appeal, the ultimate foundation of all knowledge. But this attempt failed. It turned out that the problem of light, too, can be solved only in a Copernican fashion, insofar as mechanics was unable to explain electrical and optical phenomena, but, on the contrary, had to be explained by them. This was a tortuous road marked by continual frustrations. Whenever new theories have been constructed, there appeared also new experiences accentuating the inadequacy of the solution that had been achieved.

The first and most important step toward the understanding of light was taken already at Newton's time by the Danish astronomer Olaf Roemer. It was a discovery of profound significance: in the year 1676 this astronomer determined the velocity of light and thus discovered, not only a new numerical result, but also a new physical concept. Up to that time the idea that light required time to propagate did not occur at all to anybody. Among the scholars only a few outstanding minds had foreseen the possibility of such a fact. Nowadays, when the younger generation acquires this information on the school-bench, it is taken as a matter of fact; but one should understand to what extent it contradicts immediate experience. It seems natural to us to think that light fills the room the moment we switch on an electric lamp; actually this is not at all the case, for light spreads gradually from the electric bulb and its environment to the rest of the room. The word 'gradually' is here used, of course, in a figurative sense: the process of the propagation of light takes in this case less than one-millionth of a second. This immense velocity of light was the main reason why the character of light as a spreading process could be recognized only at a late period. Only exceedingly exact measurements could determine the minute periods required for the propagation of beams of light.

This discovery remained therefore reserved for astronomy, a science combining precision of measurement with the observation of tremendous distances; it offered suitable conditions for the determination of the velocity of light. Olaf Roemer investigated the eclipses of Jupiter's satellites; he watched the disappearance and re-appearance of these moons when, in their orbital motion, they passed the cone-shaped shadow of the planet. As a result, he found that the durations of such darkenings of the moon were not always precisely the same but varied by seconds, according to the time of the year. Such little deviations from exact figures led more than once, in the history of science, to deepest insights into the nature of the world. It is as if nature discloses its fundamental relationships in the minute errors of current theories.

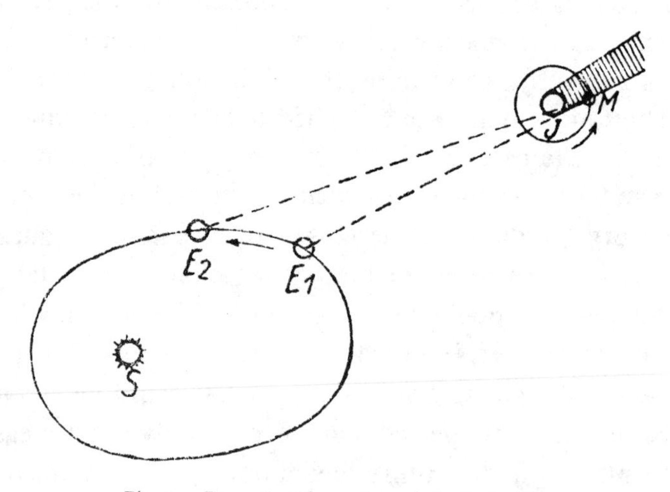

Fig. 2. Roemer's Observation of Jupiter's Moon

In Roemer's case, the existence of a velocity of light was inferred from such deviations in observations, and even the numerical value of this velocity could be calculated rather exactly. The trend of his thought can be understood, when Figure 2 is examined.

The path of the earth is here portrayed as an ellipse with the sun (S) occupying one of its foci. Jupiter (J), with the orbit of one of its moons, is found to the right (It is understood that the limitations of the diagram make it impossible for us to give a true picture of distances and sizes). When the moon enters the conical shadow of Jupiter at point M, it sends the last beam of light, reaching the earth several minutes later at point E^1. After a few days the moon emerges from the conical shadow, turns slowly around Jupiter and reaches once more point M (In reality, this is not the same point M, insofar as Jupiter with its moons will have moved forward; but this movement is very slow and can be disregarded in our explanation). At the moment of this second disappearance, the moon sends again its last beam to the earth. The latter has moved in the meantime to E^2, however, so that the beam has now a longer trip to make. Had the earth remained at E^1, the astronomer would notice the disappearance of the moon at M every time after a definite interval corresponding to the time required by the light to traverse the distance ME^1. On both occasions the delay would be the same, and the duration of a complete orbital course of the moon would be found identically correct. But the earth has not remained standing still but has moved in the meantime to E^2. Light has now a longer route ME^2 to traverse, and the excess of time required for it becomes responsible for a faulty prolongation of the orbital period. As the correct duration of each revolution of the moon is known from other sources (which cannot be here discussed), and as the distances ME^1 and ME^2 can also be estimated, the difference between the two intervals of time required for the propagation of light can be readily calculated. The time required by light to traverse a distinct distance becomes thus known, and the velocity of light can be immediately determined.

Roemer's discovery was known to Newton, whom we meet here in an important role, not only in connection with mechanics but also in that with optics. Newton explained the propagation of light as the emission of tiny particles thrown into space and capable of passing through air and gases by virtue of their smallness. He was able to account for many optical phenomena by means of this *theory of emission* of light. His doctrine dominated the physical interpretation of the world for one century, even though there was formed at that time the wave theory of light, which replaced Newton's conception at a later date.

It was the mathematician Christian Huyghens who recognized, with remarkable keenness, the possibility of explaining all phenomena of light-transmission by means of wave-propagation. His theory found acceptance in the scientific circles with considerable difficulty mainly because he put as it were the cart before the horse. It was eminently suited to explain quite simply the phenomena discerned in difficult optical experiments; but when it came to the most ordinary, easily observable facts of light-propagation, it had only extremely involved explanations to offer. Thus, it made the phenomena of the bending and interference of light easily understood; but the rectilinear propagation of light, occurring in daily experience as one of its most conspicuous characteristics (e. g. in the formation of shadows), could be conceived only as a very complicated process arising out of a peculiar superposition of light waves coming from various directions. That is why science had to cling to the emission-theory of light as long as there remained hope for Newton's theory to explain the phenomena found in experimentation, no matter in how intricate a manner. When finally, under the pressure of the results of additional experiments of great merit, the wave-theory won, it was shown that the principle, often regarded as self-evident, that 'natural' phenomena are basically 'simple', did not always hold true. Rather, it must be said today that, in general, the simplest relations in nature hardly ever appear "naturally", but must be created in laboratory conditions by means of an artificial control of active factors. The simplicity of natural processes, on the contrary, appears as an illusion due to the confluence of intricate factors. Whoever looks from a high mountain at the smooth surface of the sea, will not be inclined to think that, in reality, it has the character of a wave-like curved surface; rather, he will visualize it on a large scale and consider it as a plane. Similarly, when we face nature in everyday experience, we see it only in a broad outline. It takes the sharp eyes of science to notice behind it the intricate pattern of interconnected factors and to recognize in them the true configurations of natural forces.

The history of scientific optics is a continuous triumph of systematic methods over naive beliefs. It is easy to understand, therefore, that men outside the field of the natural sciences, whose outstanding achievements in other subjects were a result of straight-forward thinking and immediate relationship to nature, attacked again and again scientific optics for being essentially on the wrong path. Such individuals as Goethe and his various adherents failed to see that the natural sciences of the modern era arrived at their complex doctrines through a searching study of nature rather than through sheer speculation or abstraction from reality; that they can make inquiries into nature in a more exact way, because laboratory conditions permit phenomena to occur under controls which do not exist in nature; and finally, that a confident acceptance of the immediate evidence of the senses is nothing else than an uncritical overestimation of this somewhat crude set of organs, which can demonstrate its real vigor only in co-operation with keen and far-reaching powers of reason. One is tempted to remind the critic of the physical theory of colors of his own words — "if you despise reason and science, man's loftiest power." Let us leave alone, however, this quarrel over the theory of colors; it appears advisable to consider this quarrel from the standpoint of psychology rather than from that of natural science.

Facts gathered in connection with the phenomenon of interference helped a great deal to bring about the victory of the wave theory of light, absurd as it may seem to a mind guided solely by immediate experience. The substance of this theory can be described in this way: the addition of two brightnesses results in darkness, or, to use an equation:

light + light = dark

This phenomenon is not observed in daily life; it requires for adequate observation a special arrangement of light-rays. A theory considering light to be of material nature was unable to account for this equation, as a combination of two material particles can result only in more material, not less (Newton thought of explaining the phenomenon of interference by supposing that light-particles are equipped with special "fits"; but such an attempt at an explanation would presuppose essentially a compromise and must be rejected by a consistent wave-theory).

On the other hand, for the wave-theory the phenomenon of interference is obvious. Imagine a wave produced by the swinging of a rope attached to a flag-pole; the arrival of a wave-crest at the top of the pole will result in a shaking of the pole, and a similar shaking in the opposite direction will be produced by the arrival of a wave-trough. If we produce two waves in the rope in such a way that the crest of one and the trough of the other reach the top of the pole simultaneously, then the crest and the trough will cancel each other, and no tremor of the pole will occur. This can serve as an illustration of our equation; it can be written in the following form:

push + push = repose

The above equation of light can now be well understood, if we regard brightness as a push of a light-wave which is characterized by a double direction. A schematic representation of the interference of such cross-waves is given in Fig. 3.

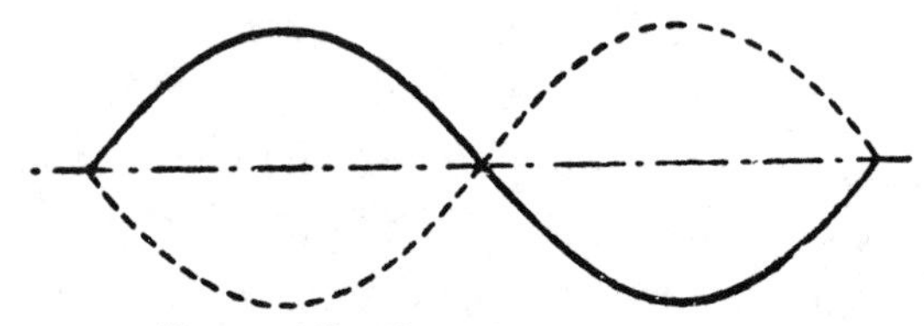

Fig. 3. The Phenomenon of Interference

The great merit of making the theory of light-waves plausible belongs to the French physicist Fresnel. He made a particular investigation of the problem of the exact nature of light-waves. There are longitudinal and transverse waves; to the latter class belong, for instance, water-waves, in which individual particles of water dance up and down and thus move transverse to the progressive direction of the wave. In longitudinal waves, on the other hand, individual particles dance back and forth in the direction of the propagation of the wave, so that a thickening and a thinning takes place as a result, and spreads forward; sound-waves exemplify this case. Fresnel was able to determine that light is connected with transverse waves, and his studies dealt primarily with the so-called polarization of light, a phenomenon characterized by the 'transverse quality' of light.

But if light has the nature of waves and is, consequently, not a substance, but a phenomenon of motion in a medium — what then is that medium itself? This is the famous question concerning ether, to which now we must give some attention. The originators of the wave-theory believed as a matter of course that the propagation of light must be conceived as a wave in a medium; and they designated this imaginary medium ether, thus availing themselves of a very old notion in natural philosophy. As a matter of fact, in all other phenomena of waves such a medium is definitely known and the necessity for it seems to be apparent. The water-wave, for instance, can come into existence only because material water particles dance up and down, so that, while each adjacent particle executes the rhythm of the movement a little later, there arises a lateral movement of the wave; this movement presents an immaterial phenomenon on a material background. Apart from such a background, wave movement appears to be unthinkable. It seems to be inseparable from the presence of matter — and this assumption is the deep source of all attempts to discover the ether of light.

However, if there is a substantial medium, it must manifest itself in other ways than in the propagation of light. We do not have to infer the existence of water from the observation of waves. There are other direct activities demonstrating to us the existence of water, such as resistence to movement or the feeling of wetness, experienced in contact with water. True enough, we should not expect such crude manifestations from ether, supposedly the finest substance permeating the pores of solid bodies. But there must be some effects demonstrating its existence; it must be possible to prove its reality by means of the finest physical instruments. In fact, the history of physics is full of most ingenious attempts to demonstrate the existence of ether and to reveal its nature. But the results, we must concede, were completely negative.

A detailed description of these experiments is out of place here, though one of them will be discussed in the next chapter. Suffice it to mention that the 'transverse' character of the light waves brings troubles in its wake, insofar as only longitudinal waves should be expected in such a fine medium. Furthermore, there arises the question of currents in ether. Similarly to water, there must arise in ether not only wave-motion but also current-motion resulting, in the vicinity of solid objects and celestial bodies, in whirlpools. The appearance of such currents should be discernible as disturbance in the propagation of light. But nothing of the kind has ever been observed. The whole mastery of optical experimentation has been used in the pursuit of some proof of the existence of ether, but all in vain: the results obtained can be accounted for only on the assumption that there is no ether.

Thus natural science found itself in a most peculiar situation. Its experiments speak against the theory of ether. What then speaks in its favor? In the last analysis, only speculative considerations compel us to accept it. However, these considerations are of extraordinarily convincing character. This is the compelling idea: if there are wave motions, there must be a medium. Thus reason is opposed to experience, and either one or the other must win in the end.

In such a conflict it is proper to subject the idea to a critical revision. There have been many ideas claiming an absolute validity and supported by the persuasive power of logical conclusions; yet they have been unable to withstand a deeper criticism. The concept of ether has not been formed on the basis of a logical conclusion, to be sure: it has an altogether different source. All common ideas comprising the knowledge of nature, such as substance, matter, wave, or motion, have not sprung out of pure speculation, but out of primary experiences of daily life. And nothing is more dangerous than to forget their origin and to ascribe to them a necessary and unconditional existence Quite on the contrary, it is important to comprehend that they have grown out of crude observations of nature, that they are hardly more than superficial generalizations concerning the world, and that it has never been demonstrated that these ideas are applicable to a finer understanding of nature.

Material substance is definitely such an idea tending to endow something highly intricate with a logically simple form. What a complicated conglomeration of matter and forces is, for instance, the substance of water! One has to think only of the atomic theory portraying it as a turmoil of individual particles attracting each other or repelling each other, sometimes mutually dependent, sometimes completely independent. A more faithful picture of the substance of water resembles a shower of bullets rather than a uniform substance. We may take it for granted that the concept of substance, characterizing this intricate picture, will do for all practical purposes. But will it do, when the explanation of the finest foundations of natural processes is at stake?

This question has to be asked, thoughtfully, only once to plant a seed of doubt in our hearts with regard to a positive answer. We should assume, on the contrary, that the concept of material substance is hardly applicable to the propagation of light, occurring both in the interspaces between the atoms and in the astronomical realm; it is a concept formed to fit the 'macroscopic' relations. If this is the case, then the natural scientists will do wisely to worry as little as possible over ether and face the possibility that there is no ether at all. In other words, there may exist an oscillating process of propagation, which is not in any sense connected with a material medium. Why should we not form this new conception conforming so much better to the experience of optics? Must we transfer, under all conditions, the 'macroscopic' ideas to 'microscopic' dimensions? May we not form, in view of highly complex and exact experiences of science, new fundamental principles doing justice to our new knowledge?

That scientific optics could and did take this path was a result of the progress made in the meantime by another physical discipline, the theory of electricity. Here we became acquainted with forces of an entirely different kind than those of mechanics familiar since earlier days; the experimental investigations of Faraday, above all, showed that, not only the electrical current flowing in the wire, but also the electric and magnetic fields found in the air or empty space, contain in reality power and energy. One thinks of magnetic and electric lines of force in terms of iron filings, as a sort of proof; these lines manifest, with a lawfulness of their own, the existence of electric and magnetic states permeating space and penetrating bodies.

It is not necessary to regard these states as states of a special substance, like that of ether; if these fields are to be considered as substance, then it is a substance of an entirely different kind from that of material bodies, such as water and air. They lack, above all, a very important quality of matter, namely, that no two bodies can occupy one and the same space—that is, impenetrability. On the other hand, two electrical fields can be superimposed without excluding each other, for the simple reason that they do not enclose any space whatsoever. It is incorrect to retort with the statement that a similar thing is observed in the mixing of fluids or gases. As a matter of fact, such a mixing should not be understood as placing the molecules 'within each other' but rather as placing them 'alongside each other', so that every one of them encloses space according to the principle of impenetrability. Two electrical fields, however, are able to occupy one and the same space at the same time, not in the sense of mixture, but as being 'within each other', whole or part; they form together a new electrical field, in which either of the two fields can be demonstrated at any time. If the electrical fields are construed as substances, then the concept of substance unavoidably acquires an entirely new meaning; so that it is clearly advisable to retain the old idea of substance and to regard the concept of 'fields' as its opposite.

We may say, then, that the study of electricity has taught us to conceive materiality in a form different from substance, namely, in that of field. To this latter concept we owe the victory over the prospectless theory of material ether.

It was the Englishman James Maxwell who took the decisive step in reducing optics to phenomena of electricity. Taking Faraday's experiments as the starting point, he sought a mathematical formulation of the fundamental principles of electricity and finally presented them in the form of the famous Maxwellian equations; the result was a concatenation, i. e. a binding together, of electric and magnetic conditions as observed in the phenomena of induction (consisting in the creation of a magnetic field by means of an electrical current, or vice versa). Maxwell noticed, however, that a mathematical development of his basic principles necessarily led to the conclusion that there must be electrical vibrations spreading through space. He immediately assumed that these vibrations must be identical with light and that light is, consequently, nothing other than an electrical phenomenon similar to the electric or magnetic fields arising in the vicinity of electrical currents; the former differs from the latter merely in the extraordinarily high rate of vibrations. He himself could give no experimental proof of this mathematical theory; the proof had to await the discovery of improved methods of observation.

The confirmation of Maxwell's theory was reached along two lines. On the one hand, it became possible to show the effect of electric and magnetic fields on light-generating structures or radiant atoms (Stark's and Zeemann's effect) and thus to prove that the emission of light is essentially an electrical phenomenon. On the other hand, long before these experiments took place, there came the great discovery of Heinrich Hertz: he succeeded in producing, by means of an electrical apparatus, electric vibrations which, though of considerably lower frequency of vibration than that of light, showed properties related to it and which could spread through space by themselves and independently of wires. These electrical vibrations produced by Heinrich Hertz in his laboratory were nothing other than wireless waves, known today as radio waves. Their widespread technical use in telegraphy and radio constitutes a proof of how a discovery made purely for theoretical reasons, that is, in search of understanding natural phenomena, can yield unsuspected industrial benefit, never thought of even by the discoverer himself.

Electrical waves are advancing fields which should not be regarded as bound to a material medium. They are waves in which electricity continually alternates between positive and negative. Yet they are not dependent on the ups and downs of small material particles, but move quite independently through space. They thus have qualities found by the science of optics in the slow course of experimentation with light. We are able to say today that light is simply a train of electrical waves of high frequency.

The pursuit of this profound knowledge has yielded us an insight of unsuspected richness into a multitude of electrical waves. We have succeeded in producing electrical waves the frequency of which is by far greater than that of light. These waves of high penetrating capacity are the X-rays, discovered by Roentgen. The examination of radioactive substances has proved that they are sending out even faster vibrating and more penetrating radia-

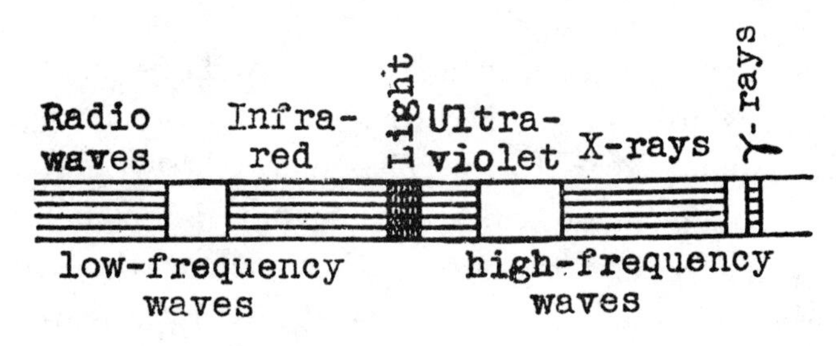

Fig. 4. The Total Spectrum

tion, namely, the gamma-rays related in many ways to the X-rays. Moreover, we have succeeded also in bridging the gap that previously existed between the light rays and the waves of the wireless, the progress having been made on both sides. On the one hand, the waves of the wireless telegraphy have been shortened (higher frequency means shorter waves); on the other hand, longer waves which no longer possess the property of being seen by the human eye, have been isolated among the light rays. The totality of these waves — the so-called spectrum — is represented in the order of their wave-lengths on Fig. 4.

Thus we have come to regard light as a rather narrow section in the whole spectrum of electrical waves. There are electrical waves of every frequency, from 0 to almost any magnitude. The highest known frequencies lie in the trillions (gamma-rays). But the human eye is sensitive only to a very small stretch of frequencies called light. The eye does not respond to the waves of other frequencies, and we need complicated apparatus to get acquainted with them.

The limitation of the eye to a definite field of frequency has its source in the history of man's development. The realm of electrical waves sent by the sun appeared to the eye as light; these rays are abundantly represented on the surface of the earth and permit an exchange of action between human beings and things, which we call 'seeing'. It cannot be called impossible that our eyes may become adjusted to other waves, for instance, to those of the wireless telegraphy; but our biological organization prevents this, insofar as we cannot change our adaptation quickly — in the manner of a receiving radio-set — so as to adjust ourselves to other waves. Consequently, we avail ourselves of physical instruments, modify the action of waves with a frequency higher or lower than that of light, and finally bring about effects which our sense organs can register as visual or auditory phenomena. However, when we visualize the whole scope of electrical waves (as represented in Fig. 4) and notice the little band of rays perceptible as light, it appears to us as if the world were covered with a curtain with a small hole through which we are allowed to contemplate only a fringe of nature's immense riches.

In conclusion, one may be desirous to raise the question: But what about sound waves? The truth is that sound waves do not enter here into consideration at all. Though they are waves, they have no place in Fig. 4: for they are not electrical waves. Rather, they are elastic vibrations in a medium, with qualities similar to those formerly ascribed to light. Their 'ether' is the air; they cannot be considered as fields. They are vibrations in a substance, not unlike the waves of water. Sounds are, therefore, inseparable from a medium. The sound of an electrical bell dies in a vacuum. In small inter-atomic regions there can be no sound, as the concept of substance, essentially macroscopic, has here no application. The sound waves, as completely macroscopic phenomena, offer us a picture of how light should not be conceived. For light, by virtue of its electrical character, stems from deeper foundations than the crude substance of the corporeal waves.

Chapter 3 : THE SPECIAL THEORY OF RELATIVITY

THE facts and considerations given in the preceding chapter led us to the conclusion that light is an electrical process rather than a mechanical one. It is not related to either water waves or sound waves. It is more akin to radio waves emitted into space from aerials and consisting in rapid changes of an electric and magnetic field. With such a statement, it is true, the problem of the existence of ether, assumed formerly, is not yet answered in the negative. All that is proved is that ether is not a substance, in the mechanical sense of the word, comparable to what we call matter. The question remains: Is it not possible that electrical prenomena may also be grounded in a substance? Can't there possibly exist a particularly fine substance underlying electrical fields and related to them as water is to water waves? Don't electrical phenomena become intelligible only when an ether is assumed?

The question of the existence of such an electrical ether cannot be dismissed without further ado. An ether may exist; yet it should be realized that the supposition has an exceedingly weak foundation. It rests on a belief which is unlikely ever to be verified, namely, on a belief that the phenomena occurring within the fine pores of matter do not appreciably differ from those occurring in the cruder material structures accessible to our senses. This conjecture is not justified by anything we know; for indeed, the progress of natural science has shown in all of its fields that nature is different, in its inner organization, from what it appears to our crude senses. Let us recall, for instance, the discoveries of biology, the science of living beings, which inform us that all living organisms consist of countless cells producing a unified living being only in collaboration. No one can say that this assumption is supported by the evidence of vision; yet it is true. And one should not be surprised that the science of physics, looking far deeper into the nature of things than biology, has come upon even greater discoveries. It seems that the vast changes in our ideas concerning the physical world are an outgrowth of the fact that the requirements of scientific precision have grown quite substantially. As long as men are satisfied with the range of exactness given by sensory perception, they can put up with a rather simple explanation of nature. But as soon as the precise measurements made possible by the modern art of experimentation are introduced, inexactitudes and contradictions are found in current theories; as a result, involved theories have to be devised to make facts agree with interpretations. Thus, the tremendous development in the field of theoretical physics during the last century was an effect of achievements of experimental physics. One should not forget that the physicists were not led to their bold assertions by mere ecstasy of speculation: they were guided by the urgent need to make theories and facts agree and to explain the discoveries revealed by improved physical instruments.

In fact, Einstein's theory of relativity, the most magnificent achievement of modern physics, was suggested by closest adherence to experimental facts; this is its strength. We may admire the grandeur of its structure of thought and the depth of its ideas; but this alone would never have secured for it that firm position in physics which it enjoys today. This position was secured because it is able to explain experimental facts, to foretell events; it was the later confirmation of these events which made this theory great.

Einstein built his theory on an extraordinary confidence in the exactitude of the art of experimentation. A number of physical experiments were under consideration, at that time, which aimed to determine the state of motion of this hypothetical light-ether. To be more exact: as ether was supposed to fill the whole of the world's space, the earth had to move through it. The goal of these experiments was to measure the motion of the earth in regard to ether. The result of all these experiments was, however, negative. The existence of ether could not be determined. It was at this point that confidence in the results of experiments became significant: Einstein was certain that the experiments would have had a positive result did ether exist at all; he concluded, therefore, that there is no such thing as ether. This conclusion as regards the non-existence of ether could be ventured only insofar as it presupposed the unconditional trustworthiness and exactness of experimental findings.

We must here describe more accurately the trend of thought which led to the decisive examination of the existence of ether. If one maintains that there is no ether, one must comprehend that such a statement requires conceptual clarification. It can mean only a definite assertion concerning the properties of light; namely, that light has no properties of the kind characterizing "coarse" waves, exemplified by waves of water or air. Among the properties of substance, in the old sense of the word, we include impenetrability; and we have shown that this property does not apply to light as an electrical field. There is a second property of substance—the determination of a state of motion. We must now clarify this point.

When we observe a water wave, we necessarily ascribe to it a certain rate of velocity. The wave takes a period of time to travel from a ship to the shore. This velocity is determined by the nature of water, by the speed with which each water particle carries along the next one, by the power of the inner cohesion of water. It is clear, moreover, that the time required by the wave to traverse a certain distance depends on one more factor. Suppose it is low tide, and water recedes away from the land; then, obviously, the period of traveling will be lengthened, for the wave will be retarded. The velocity of the wave is normally considered with regard to the water's surface. If, however, this water surface is as a whole in motion, then this motion must be added to, or substracted from, the velocity of the wave, according to its direction. The speed required by the wave to reach the shore is composed, therefore, of two velocities, that of the wave and that of the water surface. Consequently, the combined velocity will vary with the direction. In the case of a low tide, the velocity of the wave in the direction of the shore will be retarded, while the velocity of the wave moving from the ship to an island situated farther in the sea will be increased. Only with regard to the water surface is the speed of the wave equal in all directions. That is what is understood by the determination of a state of motion. If we apply measurements to water as our reference system, then there prevails an equal velocity in all directions; and the state of motion of water is, consequently, the distinctive state of motion, in terms of which the calculated velocity of the wave receives its natural value.

Such reflections were entertained with regard to ether and in connection with astronomical relations. As light traverses the world's space, ether must fill it like a great mass of water in which planets float like isles. Insofar as planets move around the sun, they must be characterized by a different state of motion from that of ether. Thus one comes to the assumption that the velocity of light, as measured on a planet like the earth, must vary with direction, simply because ether is understood as a substratum of light waves and only with regard to it can the velocity of light receive its natural value. In the eighties of the last century, an American physicist, Michelson, devised his famous experiment (since repeated many times) designed to test this line of reasoning.

The arrangement of Michelson's experiment is graphically presented in Fig. 5. The apparatus consisted of two horizontal metal bars AB and AC. In A there is a

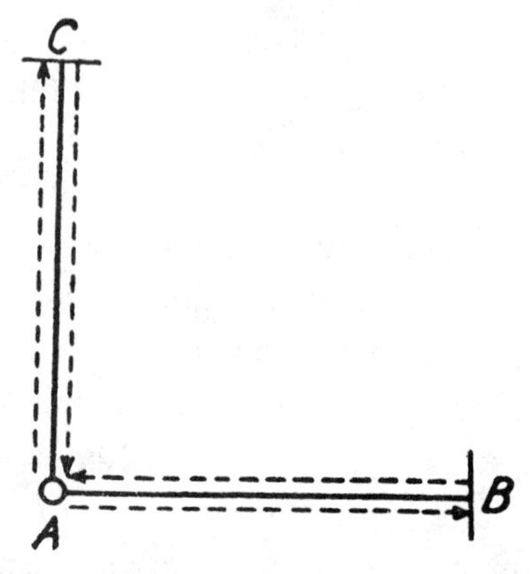

Fig. 5. The diagram of Michelson's experiment.

source of light from which rays are sent to B and C where they are reflected in a mirror and meet again at A. The dotted arrows of the figure are supposed to indicate this path; for a better view of the whole process they have been drawn partly below and partly above the bars, whereas the real path in both directions lay of course exactly in the axis of the bar. The question is: if the rays leave A simultaneously, will they return to it also simultaneously? This would be the case were the apparatus and its metal bars to rest motionless in ether, for then the speed of light is equally great in both directions AB and

AC. But the apparatus rests on the earth and hence participates in the motion of the earth through ether. It follows that the velocity of light must be different in the two directions. A simple calculation shows that, when the earth moves through ether in the direction AB, the ray A-B-A must return to the starting point a little later than the ray A-C-A.

Michelson felt sure at the time that it was possible to prove the tardy return of that ray; after all, his methods were exact enough, and he used the finest optical instruments. The belated arrival of the ray could be proved by means of interference, by the appearance of shadow-bands created by the coincidence of hills and dales of the two currents of waves (see Ch. 2). Yet the surprising result was that no shadow-bands appeared at all: there was no retardation of the ray.

This unexpected result kept the scientific world long in perplexity. The first man to attempt an explanation of the phenomenon was the Dutchman H. A. Lorentz. He assumed that the bar AB became shorter in consequence of its motion through ether; as a result the path A-B-A became shortened, and the ray came back just as quickly as the other ray. There is no objection to this explanation, except that it overlooks the fact that the problem of ether acquires a very peculiar turn. In brief, it signifies that ether exerts shortening forces upon the moving bodies in such a manner that the differences in the velocity of light connected with motion cannot be demonstrated. In other words, we are expected to believe in the existence of ether and also to assume that the proof of the existence of ether is impossible. In view of such findings, it would seem to be more plausible to stop believing in ether: for whatever defies every attempt of proof has no existence for the physicist.

Einstein accepted the latter alternative, and the convincing power of his doctrine lies precisely in its openly logical deductions. We may now formulate his view, as following from the preceding. There is no ether, in the sense of a carrying medium of light; and there is no special frame of reference in which the velocity of light is equally great in all directions. Rather, this is the case in every uniformly moving frame of reference. When measured on the moving planet of the earth, the velocity of light is identical in all directions; when measured on a differently moving planet or on a body "resting" in the solar system (such bodies, for all we know, do not exist), the velocity of light is still the same in all directions.

Einstein's doctrine signifies a definite turn in the history of the problem of ether and transforms hitherto negative findings into a positive principle. It cannot be said, to be sure, that it explains the negative findings; it proceeds the other way around and, assuming them as established, asserts that no special explanation can be here expected at all. This procedure can be compared to that of introducing the principle of the conservation of energy. Insofar as the efforts of innumerable inventors to create a perpetuum mobile have proved fruitless, this principle of energy stands for a circumscription of the fact rather than for its explanation: the feat is impossible.

Einstein's doctrine required, and was given by him, a considerable supplementation in connection with the theory of knowledge. For the contention that for every uniformly moving frame of reference the velocity of light is equal in all directions takes us in one important respect beyond the experiment of Michelson. In that experiment the velocity of light was not measured in one single direction, but as the totality of time necessary for a light-beam to travel there and back. However, how do we know that the velocity is not greater or smaller in the direction AB than it is in the direction BA, with the result that, in measuring the total time at A, the difference drops out? Is it not possible that Einstein's contention that the velocity of light is identical in both directions is a faulty hypothesis?

The answer to these questions leads to the famous doctrine of the relativity of simultaneity. This most profound of Einstein's thoughts must here be explained in greater detail.

Einstein distinguishes between simultaneity at the same spot and simultaneity of events separated by distance. This distinction becomes particularly clear when we take astronomic dimensions into consideration. An astronomical observer is attached to his spatial place; yet he receives messages or signals from distant points. He is able to record immediately only the simultaneity of their arrival to his place. Although this place is by no means a mathematical point, nevertheless it may be considered as virtually dimensionless as compared to distances traversed by light in a few seconds and referred to by the theory of relativity. The arrival of a signal may be designated as a coincidence, as a "point-event"; that is to say, as a phenomenon spatially and temporally dimensionless. Such a simultaneity at an identical point may be taken without change from the older physics. The logical problem arising beyond the realm of sensory perception is this: How does an observer arrive at the temporal order of events separated by space?

"By means of physical measurements," is the first prompt answer. The observer measures the spatial distance and divides it by the speed of the signal; thus he gets the time in which the distance was traversed. If a beam of light from Sirius reaches the earth simultaneously with a beam from the sun, then it is possible to estimate at what time each of the beams was emitted by taking into consideration the respective distances of the stars and the velocity of light.

That is, of course, correct. But first one must know the velocity of light. How can it be measured?

There is fundamentally but one method for the measurement of a signal velocity, which we shall represent schematically in the following way. Let us imagine two clocks located at two different points (Fig. 6). A signal is given at the first point, say, at 12 o'clock. It reaches the second point at 5 minutes after 12. Hence it took five minutes to cover the distance which we proceed to measure; when this is determined, the velocity in question is found by division. This is th only possible method of measuring the velocity.

But is it true? Wasn't the velocity of light measured by Michelson in an entirely different manner? Michelson sent a beam of light to a distant point and arranged for its reflection and return. He had to measure only the time at the starting point without considering the moment at which the beam reaches the mirror. However, he thus found merely the sum-total of periods necessary to tra-

Fig. 6. A Diagram of the Measurement of the Speed of Light.

verse the path to and fro. He could not determine what interests us most, the velocity in a single direction. Our contention is therefore correct.

We notice that our measurement of the velocity of light has resulted in a difficulty. In order to estimate that velocity we need two clocks at different points. In order to make the differences in time read from the clocks meaningful, the latter must be adjusted; that is to say, it is necessary to ascertain whether or not the clocks show the same figures at the same time. But we have arranged for the measurement of velocity solely for the purpose of finding a means of ascertaining the simultaneity at points located remotely from each other. We find ourselves in a vicious circle: in order to determine the simultaneity of distant events, we must know a velocity; and in order to measure the velocity, we must be capable of judging the simultaneity of events separated by distance.

Einstein has shown a way out of this logical circle: the simultaneity of distant events cannot be *verified*, it can only be *defined*. It is arbitrary; we can determine it in any manner without committing a mistake. When accordingly we make measurements, the results will contain the same simultaneity which has been introduced by definition; this process can never lead to a contradiction.

This is Einstein's famous theorem of the relativity of simultaneity. It requires a decisive change in our views, but it is unlikely that it will remain, for all times to come, as strange or bewildering as it appears to be at a first glance. As a matter of fact, anybody who grasps the idea completely will find it as intelligible and natural as the old idea of time; he will discover, moreover, that the new doctrine readily answers certain questions suppressed or neglected by the old theory. In the end he will find it difficult to think along the lines of the older view. The experience is similar to one frequently occurring when somebody goes to another country: he finds at first that he is unable to get adjusted to the new language; then forgets about it, till one day, on returning to his native land, he discovers that the new language is really more familiar to him than his native tongue.

The significance of this solution of the problem of simultaneity consists in that it makes intelligible Einstein's contention concerning the non-existence of any special frame of reference with regard to the propagation of light (and hence the non-existence of ether). Apart from this new thought, Einstein's principle would contain a logical contradiction.

This principle must now be formulated in a more exact manner. The velocity of light is identical in all directions in a uniformly moving frame of reference, provided simultaneity is correspondingly defined. This additional statement makes Einstein's contentions clear. We notice that the abandonment of the concept of macroscopic substance (together with that of a special state of motion) is bound up with the relativity of simultaneity in a peculiar manner. The profound significance for physics of investigations in the theory of knowledge thus becomes obvious.

But Einstein's theory of simultaneity has a presupposition without which it could not be maintained: it is nothing other than the assumption that no velocity greater than that of light can occur in nature. We must think it over very carefully why this assumption is so important.

For this purpose we shall explain Einstein's theory in the following manner. A light signal is sent out from A at 12 o'clock (fig. 7); it is then reflected and returns to A at 10 minutes after 12 o'clock. At what time did it reach B? According to Einstein, this cannot be determined by experiments; we can only establish it by definition. We may, for instance, record it as having occurred at 12:05; but we can think of it also as occuring at 12:02 or 12:08. But we may not declare that the arrival at B takes place at 11:59; for then the light would have arrived at B earlier than it has started from A. We know that no physical occurrences can run backward as to time. This is the only limitation; any number within the stretch of time between 12:00 and 12:10 can be chosen.

Let us therefore set the time for the arrival of the lightbeam at 12:02. Can this lead to no contradiction? There would always be a possibility of contradiction were there signals faster than light in existence. Let us suppose that there is a signal requiring three minutes less than light to traverse the distance AB. Let this signal be sent from the point A simultaneously with the light-beam. As the light-beam arrives at B at 12:02, the other signal will arrive, according to our assumption, at 12:02 minus 3 minutes, that is, at 11:59. Now, both signals were sent out from A at 12 o'clock. It follows, absurdly enough, that the new signal arrives at B sooner than it starts from A. The determination of simultaneity has led us to a contradiction; but only because we have accepted the possibility of the existence of signals traveling faster than light.

A contradiction in Einstein's theory of simultaneity is impossible only if there are no signals traveling faster than light. That is another contention of Einstein. Indeed, it is the most important contention of his special theory of relativity. The statement must be made still clearer, if we are to accept it fully.

We must admit, of course, that no physicist has up to now found signals traveling faster than light; but are we certain that such signals do not exist? There are many things, no doubt, of which we have no knowledge today, but which we may come across perhaps tomorrow. Who would have thought 150 years ago that one could travel from New York to Boston in 5 hours, a distance requiring at that time at least several days? Who would have believed then that it might become possible to converse orally across that distance, as it is now done every day over the telephone? May not similar surprises await us in the science of physics? May not some day a spreading process be discovered in comparison to which the velocity of light will appear like Stephenson's first train as compared with a modern express train?

Fig. 7. A Diagram of the Course of a Light-Signal.

Ready as the physicist may be to admit the possibility of any technical dream of the future, he cannot accept this dream. If a utopian poet should portray the day when a regular traffic to Mars began or when the highly progressed humanity rescued the earth from the chains of the sun grown cold and steered the planet toward other stars, the physicist would have no objection, for physical reasons, to such conjectures. But to every fancy in which even the smallest action spreads quicker than light, in which waves of some kind "run ahead of light" as it were, he must respond with a blunt "impossible." Cautious as he may be in denying possibilities, he realizes that there are denials which must be uttered with assurance, unless his entire science is to lose its meaning. There are denials expressing a law of nature; and this is one of them.

Such denials are, after all, common in physics. One can easily show that every law of nature carries within itself a statement of denial. The law of the conservation of energy, for instance, can be expressed in this form: there will never be found a process, even in one hundred thousand years, in which the amount of energy increases apart from an outside influence. Thus, the positive law of the conservation of energy contains within itself a negative consequence. And vice versa, the negative law of the limitation of the velocity of light can be formulated to show its positive kernel. We now want to bring out this kernel.

In the first place, Einstein brings into the picture a peculiar contention concerning the energy of moving bodies. Every body in motion carries within itself an amount of energy which increases with the velocity of the body. This energy is required to start the motion; we recognize it, on the other hand, in the impact provided by a moving body to one standing still. According to Einstein, the content of energy in a moving body grows with an increasing speed faster than assumed by the old theory. In order to bring a body up to the velocity of light, an infinite amount of energy would be required. It is therefore impossible for a body to move quicker than light; in fact, no material object can reach that velocity.

In the second place, the law of the limitation of light-velocity rests upon the knowledge that light does not constitute a physical phenomenon of its own but rather represents a special case of the transfer of electrical activity in general. In the preceding chapter we had an opportunity to see that light is an electrical phenomenon and that light waves represent only a section of the great realm of electrical waves. What is maintained by Einstein with regard to light goes, therefore, for all electrical waves of which light is but a representative. But according to our knowledge of the internal structure of all substances, there are basically only two ways of transfering power from body to body: gravitation and the electrical wave. Every other manifestation of force is composed of them. If they both move with the velocity of light, as Einstein contends, then a slowing up may occur within the atoms of the body, when the power runs in a zig-zag course; but it can never be accelerated. Einstein's law of the limit character of the light-velocity means thus nothing other than a formulation of the fact that light represents one original form of the transfer of action, the other representing an equal speed limit.

Only with the addition of this idea does Einstein's theory of the relativity of simultaneity become intelligible. It even leads to a clarification of the concept of simultaneity itself. What do we mean when we speak of simultaneity? Let us take an example. Let us say that I wish to visit a friend of mine in Southampton. I depart in a steamer from New York at 12 o'clock. Now it happens that my friend leaves Southampton for New York precisely at the same time. Neither of us knows about the other's departure. Only at the last moment do we send telegrams to each other. We shall now consider a small delay of the telegram due to its being written out and carried out, and we shall assume that the telegram arrives within a few minutes. Such a telegram is then the quickest practical signal, although the delay makes it a little slower than the velocity of light. If both telegrams start out simultaneously, each will reach its destination slightly late, that is, after the ship's departure. Had my friend left but a few minutes later, my telegram would have reached him and kept him in Southampton. And vice versa, had I left a little later, I would have received the telegram and could have avoided a superfluous trip. The fact that we both left simultaneously simply means that it was impossible either for my telegram to reach him or for his telegram to reach me. We find that simultaneity means an exclusion of causal connection. When two events P and Q take place simultaneously, there is no possible effect of P on Q or of Q on P.

If this is the definition of the concept of simultaneity, then the indeterminacy of simultaneity is at once apparent. As my telegram takes several minutes to reach Southampton, my friend could have left at 12:01 without receiving the telegram. On the basis of this "telegraphic" speed, the two events could have been called simultaneous. Now it is true that the velocity of light is considerably greater; the light-signal—or what is the same: the radio waves apart from the delay by writing and delivering the telegram—require only a fraction of a second to traverse the distance over the ocean. But light does not travel infinitely fast. Because of the great velocity of light the interval of time within which simultaneity is arbitrary is short; but it is not a nought. We understand now how the relativity of simultaneity is connected with the limit character of the velocity of light: as there is a finite limit to all velocities transferring action, a possible causal connection of two distant events is necessarily excluded for a short duration; the arbitrariness of simultaneity lies precisely within this duration.

The unique position which light occupies in the theory of relativity may be expressed also in a different manner. Whereas in Einstein's original theory of relativity light served merely to determine simultaneity, it became clear in the later revision of the theory that light may be used for all measurements of time, for the designation of the measure of time, and even for the measurement of space. One may construct a geometry of light* in which light determines the comparison of spatial distances. Thus light comes to serve as the ordering net of physics, which gathers within the meshes of its rays all the events of the world and puts them in a numerical order.

With this idea in mind, one may further represent the content of Einstein's theory of space-time in the following way. Clocks and yardsticks, the material instruments for measuring space and time, have only a subordinate function. They adjust themselves to the geometry of light and obey all the laws which light furnishes for the com-

*See H. Reichenbach, *The Philosophy of Space and Time,* English translation, Maria Reichenbach and John Freund, Dover Publications, Inc., New York, 1957.

parison of magnitudes. One is reminded of a magnetic needle adjusting itself to the field of magnetic forces, but not choosing its direction independently. Clocks and yardsticks, too, have no independent magnitude; rather, they adjust themselves to the metric field of space, the structure of which manifests itself most clearly in the rays of light.

In view of the preceding argument, this seems to be a fairly plausible statement; yet it leads to a noteworthy conclusion concerning the behavior of clocks. According to it, it is possible to show that moving clocks behave differently from those in repose. Movement exerts a retarding influence upon clocks. If a clock is moved from place to place and finally returned to its original place, it is slower than a clock which remained motionless at one and the same spot. The contention would be totally inconsequential, to be sure, were it applicable merely to clocks: the physicist then would calculate the influence of the motion and accordingly set the clock properly. But the theory of relativity maintains much more; it maintains, namely, that any running mechanism, regardless of kind, would manifest a similar retardation. Were an observer to make a journey with the clock and try to check the retardation of the clock by means of measuring devices taken along, he would be unable to notice any difference, insofar as the clock would go without any change with regard to his devices. Even if he investigates the processes of his own organism, estimates the period between two meals on the basis of hunger pangs, or measures the duration of normal sleep by the clock brought along, he still would be unable to discern any difference from previous experiences.

If this is to be fully understood, we must realize that all the processes of the human body are rooted in physicochemical changes and ultimately rest on the motion of atoms and electrons. But the processes of these elementary particles will be slowed down in the same proportion as the clock; man's feelings and perceptions will be, consequently, in complete accord with the clock.

These reflections lead the theory of relativity to assert that nobody can be forced to acknowledge the retardation of a moving clock as long as it is compared with other objects participating in its motion. One may simply declare that nothing has changed during the motion. Only regarding objects of another state of motion can we speak of a delay of our clock.

In application to astronomical relations, that is, to great distances and great velocities, these considerations lead to remarkable conclusions. Let us suppose that the above mentioned ship of space to Mars has been actually invented and that one of twin brothers undertakes the long voyage while the other remains on the earth. Years pass, and the twin at home has grown old. Then one day the ship of space returns with his brother who looks only a few years older than on the day of his departure. The brother has not noticed during his trip, of course, the fact of his preserved youth, as all of his fellow-travelers have remained in the same age relationship as himself, and all the clocks on board have made as many double turns as there have been days of the travelers' aging. Subjectively, the traveler lived but a few years, while the persons remaining on the earth lived through a great many years. If the traveler remains on the earth, the period of his whole life, from his own standpoint, will appear to him no longer than that of other people; but now he will be able to reach a much later age than his brother and his generation of men will ever be able to attain.

This example has caused much surprise and even controversy in the discussion of the theory of relativity; but it is impossible to deny that it follows necessarily from the theory of relativity and that all physical facts speak for the correctness of the contention. The theory of relativity will not declare, to be sure, anything concerning the possibility of ever traveling across the space of the universe, for the simple reason that prophesies with regard to technical progress are outside its domain. But it may assert that, if such a trip is ever undertaken, the travelers are bound to age slower, as explained in the above example. The hypothetical form of the assertion is right, even compulsory, insofar as all available facts are in favor of the doctrine of relativity. We cannot accept the objection that the case is inconceivable. Quite the contrary, everything described in it is quite conceivable; and fiction has more than once resorted to such imagery, for instance, in the form of the monk of Heisterbach. The novelty of the case consists only in that it is now the imagery which represents the truth.

Since we have undertaken to illustrate the contentions of the theory of relativity by cases of astronomical utopias, let us add one more remark concerning celestial telephoning. Our statement to the effect that no signal can travel faster than light leads to rather sad conclusions, in this connection. A beam of light requires about 8 minutes to cover the distance separating the earth from the sun, and 16 minutes to cover it both ways. The distance of Mars is sometimes greater, sometimes smaller than that of the sun, and therefore the corresponding figures will vary. Let us take an average position of Mars, the distance of which corresponds approximately to that of the sun; in that case, the electrical waves conveying a telephone conversation will take 16 minutes for the round trip. This would mean that, in making a call to an inhabitant of Mars, we must wait a quarter of an hour to get an answer to a question. Such slowness of communication would be quite unpleasant, and the cozy chats characteristic of everyday telephone calls would hardly occur in communication with Mars. The situation is considerably worse with regard to fixed stars and their planets. In fact, the nearest fixed star is about 8 light years away from us. We would have to wait at the telephone receiver for sixteen years to get an answer, not to mention the case of more distant stars an answer from which could be received only by our great-grand-children.

The prospects for celestial intercourse compare unfavorably to those of traveling. There is no limit to possibilities of reaching remote planets. One might surmise

Chapter 4 : THE RELATIVITY OF MOTION

THE idea of the relativity of motion, which gave Einstein's theory its name, leads us back to the older root of this theory, referred to in the first chapter. The Copernican view of the world and its consolidation through the mechanics of Newton have become the starting point of reflections which began to bear fruit only after Einstein combined them with his criticism of the problem of ether. To be able to understand this, we must examine somewhat closer the problem of the relativity of motion.

The idea of the relativity of motion has a strangely compelling force, once it is well understood. Who is not familiar with the phenomenon commonly experienced in a railroad car: one's own train stands still, while a train on the next track starts moving—but the impression is opposite, that one's own train has started. Only after a while does one notice the illusion. But a thought may occur in connection with this experience: what right have I to call what I distinctly saw an illusion. Was it an illusion? Was it untrue? May I not contend with an equal right that the other train stood still while my train was moving? To be sure, I had not noticed at the time that the surroundings, e.g., the depot, remained standing still and that I, therefore, was motionless with regard to this environment. But what of it if I include this environment into my conception? May I not then declare that the other train stood still and that my train together with the depot, even the whole earth, was moving past it? May I not declare this with an equal right?

Once this idea is understood, it is impossible to get rid of it. It is easy to see that the large size of the depot, as opposed to that of the moving train, cannot serve as a disproof: the difference in size is quite irrelevant. If two bodies located in empty space, a large one and a small one, were to move toward each other, should one say that the large body is standing still while the small one is moving? This would make no sense. That motion cannot depend on size is clear from a situation in which the bodies are of equal size; here size certainly cannot determine which body is at rest.

The following consideration holds true. Suppose that body A is at rest and body B is moving toward it; the movement would be recognized by the diminution of the mutual distance. Let us then suppose that B is at rest while A is moving; again we notice only the diminution of the mutual distance. There is, therefore, no way of concluding from the observed phenomena as to which of the bodies is moving, insofar as the observed phenomena are the same in both instances. Hence it is nonsensical to speak of a "true" movement. One can only say that the bodies move toward each other; their movement is relative. This is consequently the answer toward which such a process of reasoning leads: there is no true movement, no absolute movement, but only relative movement.

This idea has been repeatedly uttered. And it is interesting that it precipitated once before a quarrel over the relativity of movement, a quarrel which received then no less publicity than Einstein's theory in our days. It happened at the time of Newton and Leibniz; Newton's theory of absolute motion was combatted by Leibniz. The famous correspondence, in which these questions are discussed, has been preserved since those days. Leibniz defended in it the relativity of motion against the theologian Clarke, a friend of Newton, and offered for his views arguments which even today play a part in the discussion of relativity. He emphatically stated that all appearances are the same, regardless of whether one ascribes motion to one or the other of the two bodies. The problem, he added, is not different even in the case of one thousand bodies, and "the angels themselves" could not decide, on the basis of the observed phenomena, which body is really in motion. From Leibniz comes also the demonstration of the concept of relativity by means of the famous principle of the identity of indiscernibles; what is indiscernible is not different, and it is therefore meaningless to talk of absolute motion.

Nevertheless, the grounds cited by Newton in favor of absolute motion could not be weakened by Leibniz. Newton realized that all familiar proofs of the relativity of motion can be justified only kinematically, that is to say, insofar as motion is regarded as a change of place, as a visible phenomenon requiring no reasons. But the moment one starts looking for the active forces of motion the picture changes completely; and therefore, points out Newton, the relativity of motion is untenable dynamically, that is, from the standpoint of the theory of forces. To understand this we must give an outline of Newton's theory.

First of all, Newton differentiates between uniform and accelerated motion. A body left by itself in an empty space will not change its motion; it will move at an even speed and in a straight path. To the law of inertia, already established by Galileo, Newton added this thought: there is a force responsible for every change of motion; and conversely, the presence of forces indicates that the body is not in a uniform, but an accelerated motion.

The same reasoning applies, correspondingly, to a retarded motion. It has become therefore customary in science to regard the retarded motion as "negatively accelerated." This is merely a convenient method of expression, which no one need abhor. The circular or "rotary" motion is also considered as an accelerated motion; though its velocity may remain the same as to magnitude, it continuously changes its direction and consequently cannot be classified as a uniform motion.

The rotary motion offers an excellent illustration of Newton's idea of the absolute motion. Let us take an example. Imagine a merry-go-round surrounded by a round building similar to what we see at fairs. When we sit in it, we get fairly soon the impression that we stand still, together with the merry-go-round, while the building moves around us. If we forget for a moment what we saw before getting in, namely, that the building stands firmly on the ground and that the merry-go-round is equipped with wheels, have we any way of determining, while sitting in the merry-go-round, whether it is the building or the merry-go-round that moves?

Indeed, we have. For we feel, while sitting in the merry-go-round, an outward pull caused by the so-called centrifugal power. This power forces us against the railing. Were the merry-go-round to stand still and the building to move, then the sight for the eyes would be the same, but the push toward the railing, the centrifugal power, would not be there. A true state of rest can be recognized by the absence of the centrifugal power. Its appearance or disappearance plays a decisive role in the question of absolute motion.

This was Newton's idea explained by him in a similar example (that of a revolving pail). We can, he declared, determine even the direction of the rotation. Suppose there is another, smaller merry-go-round attached to the larger one approximately at its center, but revolving in the opposite direction. We climb now into the smaller merry-go-round and investigate: is the outward push (that is, the centrifugal power) stronger or weaker than in the larger one? If it is stronger, then the rotation of the smaller merry-go-round is faster than that of the larger one; and the direction of the rotation is the same. But if it is weaker, then the smaller merry-go-round rotates backward, in the opposite direction to that of the larger one.

We must admire the logical accuracy with which the great physicist constructed his doctrine of the absolute motion and of the absolute space. In the following lines we cite from his principal work the passages recapitulating his theory. He writes in *The Mathematical Principles of Natural Philosophy*:

"II. Absolute space, in its own nature, without regard to any thing external, remains always similar and immovable.

"Relative space is a measure of this space or a certain movable part of it, which is defined by our senses by its position with regard to bodies, and is usually taken for motionless space. . . .

"IV. Absolute motion is the translation of a body from one absolute place into another; and relative motion, the translation from one relative place into another. . . .

"And so, instead of absolute place and motion, we use relative ones . . . in philosophical discussion, we ought to abstract from our senses. . . For it may be that there is no body really at rest, to which the places and motions of others may be referred. . . .

"The effects which distinguish absolute from relative motion are the forces of receding from the axis of circular motion. For there are no such forces in a circular motion purely relative, but in a true and absolute circular motion they are greater or less, according to the quantity of the motion."

The words with which he closes the introduction to his main work show how sure Newton felt of his affirmation of absolute motion, namely:

"How we are to obtain the true motions from their causes, effects, and apparent differences, and the converse, namely, to derive the causes and effects from the true or apparent motions, shall be explained more at large in the following treatise. For to this end it was that I composed it."

These words of Newton demonstrate sharply the contrast which may exist between the objective importance of a discovery and the subjective significance attributed to it by its author. Whereas the physical work of Newtonian dynamics has become a firmly established part of science—merely raised by its later development to a higher form of knowledge, but otherwise remaining, as an approximation, permanently valid—Newton's philosophical interpretation of his work has been of a restricted duration. Nevertheless, a consistent development of the theory of absoluteness has contributed to the deeper insights of today; for only the compulsion to refute Newton's arguments could lead to the final clarification of the idea of general relativity, which was to be extended from relativistic kinematics to relativistic dynamics.

Almost 200 years had to pass before a real refutation of Newton's thought was found. In the eighties of the last century, Ernst Mach, in criticizing Newton's work, found the counter-argument. If we return to our example of a merry-go-round, this was Mach's idea: Newton has overlooked that the case of the merry-go-round at rest and of the building in rotation does not represent the opposite of the original case. He has forgotten to take into consideration the surroundings of the building, the earth, the whole universe. For, in revolving, the merry-go-round does not revolve with regard to the building alone but also with regard to the earth. In the contrary case we must let not only the building revolve round the resting merry-go-round, but also the earth and the universe—only then shall we present an equivalent but reverse picture.

But in that case, continued Mach, the centrifugal force will appear again in the merry-go-round, for this case is no other than the original one, though presenting a kinematically different description. In this description, the centrifugal force should be understood as an effect of the revolving earth-mass or even of the star-mass. These moving masses produce a pulling field experienced by me within the merry-go-round. In a quite surprising way, the concept of force becomes thus involved in the reversion leading to the two equivalent interpretations. The same observable effect, namely, the pressure against the railing, appears in one conception as a consequence of the merry-go-round's movement, in the other, as a consequence of the rotation of the surrounding masses. That rotating masses should form such a field of radially divergent forces, is for the science of physics a new but not an unusual thought. According to this conception, the Newtonian attraction of masses would be supplemented by the new forces arising out of rotary movement. One could imagine (according to Mach) that the walls of the building are several miles thick; then, in rotating around the merry-go-round, the mass of the walls would produce in the middle of the merry-go-round a field of radially divergent forces, corresponding to the centrifugal field. This field, of course, would be by far inferior in strength to that produced by the rotating universe.

Could this be demonstrated experimentally? But, remarks Mach, the proof is already available. For we do observe the centrifugal force; if we interpret it as an effect of the revolving masses of stars, then this is all that can be asked for from observation. The new conception differs from the old one only in the interpretation, not in what can be observed by the senses. Nevertheless, it may be possible to devise experiments in which the idea of Mach would lead to new observations. Imagine a rotating fly-wheel of a huge machine; it represents a rotating mass and should exercise in its interior a propelling action creating near its axis an area of "centrifugal force." Mach did not, of course, mean here the action of the wheel's own centrifugal force, from whose explosive effect the wheel is protected only by its solidity; rather, he wanted to say that a small body at rest, if placed near the axis, would be subjected to a pull toward the edge of the wheel. This action is, to be sure, so minute that it cannot be demonstrated; the mass of the largest fly-wheel is, indeed, exceedingly small in comparison to that of the universe or of the fixed stars the rotation of which produces the ordinary centrifugal force.

But even more important than this physical consequence is the relativization of the concept of force, as expressed by Mach. For, what Mach says is that in accordance with varying descriptions of the state of motion, the field of forces, too, must be presented in a different fashion. No sooner does the concept of force partake of relativity than the dynamic distinction of one state of motion disappears; and then there is no absolute motion in any sense.

Here lies the weight of the argument. The relativity of motion is tenable not only kinematically but also dynamically, if the relativization of the concept of force is introduced. Even forces are not absolute quantities; they depend upon the system of reference. When one passes to a differently moving system, the forces have to be measured differently. What appears as action of inertia when the merry-go-round is conceived as moving, appears as action of gravitation, when it is imagined as standing still and the earth as rotating. Even the Copernican world-view appears to be shaken by this consideration. It makes no sense, accordingly, to speak of a difference in truth between Copernicus and Ptolemy: both conceptions are equally permissible descriptions. What has been considered as the greatest discovery of occidental wisdom, as opposed to that of antiquity, is questioned as to its truth-value. Though this fact clearly warns us to be wary in the formulation and evaluation of scientific results, nevertheless it by no means signifies a step backward in the progress of history. The doctrine of relativity does not assert that Ptolemy's view is correct; it rather contests the absolute meaning of either view. This new insight could be gained only because the historical development went through both conceptions, because the replacement of the Ptolemaic world-view by the Copernican world-view established the new mechanics which finally provided the physicist with a means of recognizing the one-sidedness of the Copernican world-view itself. The road to truth followed here the three dialectical steps which Hegel regarded as necessary for all historical development, the steps leading from a thesis over an antithesis to a higher synthesis.

It would be saying too much to regard the fulfillment of the third stage as given in Mach's idea. When Mach replied to Newton that the centrifugal force must be accounted for in terms of the relative motion alone, he offered merely a program, not a physical theory; in fact, it was merely a beginning of a program for the physical theory elaborating the idea. Indeed, not only the centrifugal force but all mechanical phenomena must be accounted for in terms of the relative motion; the question is, above all, how to explain relativistically the phenomena of motion in the field of gravitation, i.e., the planets' movements.

It was the great achievement of Newtonian mechanics that it provided the Copernican world-view with a dynamic foundation. Whereas there existed no difference, from the kinematic standpoint, between the Copernican and the Ptolemaic systems, Newton, taking the standpoint of dynamics, decided in favor of Copernicus. For his theory of gravitational force offered to the latter view a mechanical explanation; whereas the complicated planetary orbits of Ptolemy did not fit into any explanation. If the question is how to provide both conceptions of the universe with an equal justification in terms of dynamics, then a general theory of gravitation has to be found, which explains the Ptolemaic as well as the Copernican planetary motion as a phenomenon of gravitation. Here lies the great mathematico-physical achievement of Einstein, in comparison to which Mach's thought appears merely as a first suggestion. Einstein has indeed found a comprehensive theory of gravitation, and only because of this discovery, which places his name in the same category with Copernicus and Newton, can we say that the problem of the relativity of motion has been brought, physically, to its conclusion.

Chapter 5 : GENERAL THEORY OF RELATIVITY

EVEN though the basic ideas leading to the general theory of relativity were clear to Einstein, the road to the complete theory was still long and laborious. Already in 1906, merely a year after the formulation of the special theory of relativity, Einstein had expressed the basic ideas of the new doctrine, going substantially beyond Mach. But the construction of the theory placed him before unsuspected mathematical difficulties. There was one period, in this path, when Einstein thought he had demonstrated the impossibility of a general theory of relativity. Only in 1915 did he succeed in completing the theory combining Mach's idea of the relativity of motion with the special theory of relativity into a completely new theory of gravitation, bringing thereby to a magnificent conclusion the era of classical physics. The news of Einstein's theory reached the public only in 1919, when an English expedition sent to observe an eclipse of the sun reported the first astronomical confirmation of his predictions.

In attempting to present Einstein's theory of gravitation, we must first get acquainted with the modification given by Einstein to Mach's idea. The idea of the relativity of force if stated in the form given by Mach, can be used only in connection with rotary motion. Einstein had to extend the idea in such a manner as to make it applicable to every motion. He achieved his aim through the so-called principle of equivalence.

We can clarify this principle by means of the so-called "box experiment" invented by Einstein in order to illustrate his ideas. Let us imagine a closed box of the size of a room, in which a physicist finds himself (Fig. 8). There is a spiral spring hanging down from the ceiling, to which an iron weight *m* is attached. The physicist

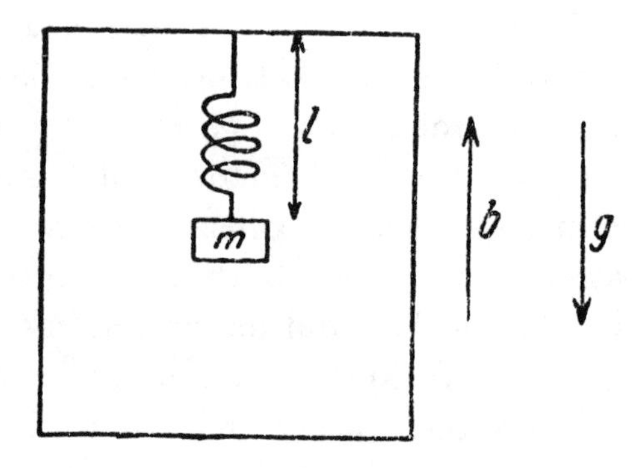

Fig. 8. *Einstein's "Box Experiment"*

has taken the measurement of the distance of the weight from the ceiling, i.e. of the distance to which the tension of the spring is adjusted.

The box has no windows. Were the box set in motion from outside, would the physicist notice the fact? Suppose that the box is being pulled up by a rope, like an elevator, in the direction of arrow *b*. Would the physicist inside notice it? Indeed he would be able to notice the change in the interior of the box: the weight m would remain slightly behind the motion, on account of its inertia; the length of the spring would increase a little, accompanied by an increase in its tension. An accelerated or growing movement would thus result in a lengthening of the spring.*

Now, says Einstein, let us assume the physicist is aware of the lengthening of the spring; this is all that he observes immediately. Must he infer a motion of the box? Certainly, he can make this inference, for the motion of the box would produce this effect; but can this effect arise in no other way? If such a second cause is possible there is no necessity to infer a motion of the box.

Now, there exists indeed a second cause that could produce the same effect. If we assume that a great planetary mass is being gathered underneath the box, then it would produce a gravitational field. This field would act on the weight in the direction of the arrow g and pull it down. Again the physicist would observe an increase in the tension of the spring as well as an increase of its length l. From the observed lengthening of the spring the physicist, therefore, could just as well infer a field of gravitation below the box, as a movement of the box upward.

But is there no way of distinguishing between these two possibilities? Are there no other experiments enabling us to differentiate between a gravitational field and

*Were the motion uniform, that is, were the velocity of the box changeless, no expansion of the spring would take place. We must, therefore, keep steadily in mind, here and in the following, that the motion of the box is accelerated.

an accelerated motion? Certainly not, as long as the physicist performs his experiments within the box. But if he cuts a window into the box and observes the outer surroundings, could he not then easily determine what is happening outside?

He could easily observe with his own eyes, indeed, whether or not any planetary masses have gathered underneath the box. We must remember here, however, that, according to the considerations given in the preceding, there are other possible gravitational fields than those produced by masses at rest; namely, the movement of masses, too, can produce a field of gravitation, called by us a dynamic field of gravitation. Let us assume that there are no masses gathered underneath the box, but that the physicist observes an accelerated motion of the box with regard to the surrounding world; must he now say that the box is in motion and the world at rest? We have made it clear already that a mere observation with the naked eye cannot inform him of anything in this connection, because it informs him only of a change in relative distances. Now we find that mechanical experiments within the box are not decisive, if the physicist takes only dynamic fields of gravitation into consideration. The physicist could then account for the lengthening of the spring in two ways:

1. The box moves upward with acceleration, in the direction *b*; the weight *m* remains behind on account of inertia; the spring becomes tense.

2. The box stands still, but the surrounding masses

move downward with acceleration; they produce a dynamic gravitational field g; the weight m is pulled down because of its heaviness; the spring becomes tense.

Both explanations are justified; there is no objective discrimination between them.

If we take a closer look at the two formulations, we notice that, whereas the last sentences sound alike, there is a peculiar difference in the sentences before the last. In the first formulation it is stated: "the weight m remains behind on account of inertia"; in the second formulation: "the weight m is pulled down because of its heaviness." Two entirely different properties of bodies, inertia and heaviness, are placed here parallel to each other. It is maintained that either of the properties leads to the same effect, namely, to the increased tension of the spring. What are the grounds therefor?

In order to understand this, we must re-examine these properties in a greater detail. For the layman does not quite know what is to be understood by the concepts of inertia and heaviness. Hence let us start with a distinction that underlies them, namely, with the distinction between mass and weight.

If we put a block of iron on the hand, we feel a pressure arising from its weight. Two factors are involved in the weight of a body: first, the mass of the body itself, and second, the mass of the earth. This double effect of active factors can be made readily intelligible in the following manner. If we take a larger block of iron, we increase the mass of the body, and thus the pressure on the hand grows. One cause of the pressure is therefore contained in the bodily mass. We can increase the pressure also in a different way, without changing the body itself. If we visit one of those places of the earth, where the gravitation of the earth is stronger, then the body's attraction is magnified and its pressure on the hand is greater. In fact, there are such places. One could, for instance, descend into a deep mine pit; or one could go to the vicinity of a pole of the earth, which lies closer to the center of the earth, on account of its flattened shape, than do the middle or tropical zones. The variations of gravitation are not, to be sure, very considerable: they cannot be felt by the hand; more sensitive scales would have to be used. The scales in question could not be of the balance type, for the weights placed in one side would increase in weight just as much as the block of iron, with the result that the scales would indicate the same weight as before. One would have to use a spring scale, similar to those used in households; then, in the places located closer to the center of the earth, the spring will be more compressed.

The weight of a body is, therefore, different from its mass; it is the effect of attraction of this mass by the earth. At a great distance from the earth and other heavenly bodies, the weight of a body would be nil, while its mass would remain unchanged. On a large planet, such as Jupiter, all bodies are considerably heavier than on the earth. Our muscular strength would not be sufficient there, for instance, to lift a child from the ground, while on a small heavenly body, such as the moon, we could pick up a grown-up person with great facility. We may define the mass, therefore, as that quality of a body, which determines its weight in a given gravitational field; the weight itself depends on that gravitational field.

The mass, if understood in this way, characterizes the body only with reference to the gravitational field and, therefore, in a rather one-sided manner. We shall call it "the heavy mass" of the body. Besides, there exists an entirely different effect of the mass, which leads us to the concept of "the inert mass."

Let us imagine a loaded railroad car. In order to set it in motion, a great force is required. This force is not directed, however, against gravitation, as the car rolls on horizontal tracks. It is the inertia of the load that opposes the motion. The applied force is, therefore, entirely independent of gravitation. In order to move the wagon on Jupiter, no more force would be required than on the earth, and vice versa; nor would this movement be easier on the moon. We designate as "the inert mass" that property which is determined by the opposition to changes in motion.

It is a fact of experience that the inert mass of a body equals its heavy mass. This is by no means a matter of course. This fact can be illustrated in the following manner.

Suppose that a log of wood and a block of iron lie on the large scales, and the two are found to be of equal weight. The log of wood is, of course, much larger. Now, both things are delivered, one after the other, to a railroad car; then we investigate whether it is equally difficult to set them in motion along the horizontal tracks. This is not a matter of course; one could surmise that the great wooden log would show more inertia-resistance than the small iron block, for their weight, or their pressure on the understructure, does not enter here into consideration. But experience instructs us that there is no difference at all. Bodies of equal weight have the same inertia; the heavy mass equals the inert mass.

This result also explains the fact that, with the elimination of air resistance in the vacuum, all bodies fall equally fast. The heavier body has a stronger downward pull, but at the same time it has to carry a greater inert mass; that is why it does not come down quicker.

After these considerations, we may return to our starting point, the physicist in the box, who is in possession of two equally justifiable explanations of the meaning of his findings. The connection of this Einsteinian consideration with Mach's criticism of the problem of rotation becomes now clear. Here, too, we find the duality of explanations: the observed effect of forces is either due to the resistance of inertia or to an overflow of a dynamic gravitational field. Whereas the observed effect was, in Mach's case, the centrifugal force and the pressure against the railing of the merry-go-round, in Einstein's case of the box experiment it is the tension of the spring, and the lengthening of l. But now we recognize the advantage of Einstein's presentation: it allows us to discover the reason for the

phenomena are to be included under the general theory of gravitation, that gravitation plays the same role in the doctrine of electricity, of optics, etc., as in mechanics.

I say that this is a truly Einsteinian turn. The physical depth of Einstein's ideas can be, indeed, comprehended only when one realizes how this method of reasoning is employed in his basic assumptions. This was the case in the special theory of relativity. It was known that several important attempts failed to confirm the existence of ether; Einstein concluded from this that, in general, no similar attempt can do better, no matter what means are used. The principle of equivalence reveals the same attitude. It is known that mechanical phenomena manifest no distinction between accelerated motion and gravitational field; Einstein concludes that this applies equally to all other phenomena. From the standpoint of logic, one cannot speak here of an inference, for this far-reaching assumption cannot be logically demonstrated by means of the scantily available facts. Rather, we have here a typical procedure in physics, that of the formation of a hypothesis; although a more extended assumption cannot be logically justified, nevertheless it is made in the spirit of a conjecture. There seems to exist something like an instinct for the hidden intentions of nature; and whoever possesses this instinct, takes the spade to the right place where gold is hidden, and thus arrives at deep scientific insights. It must be said that Einstein possesses this instinct to the highest degree. His assumptions cannot be justified in a purely logical way; yet they introduce new ideas quite in the right place. That the place is right, can be readily recognized when gold lies in front of us. In physics, too, there is subsequent justification; for it is possible to perform experiments which later verify the new hypotheses. Thus it is possible to perform experiments testing Einstein's assumption that the electrical and optical phenomena are affected by gravitation. Such experiments have been made, and they have confirmed Einstein's hypothesis in a decisive way.

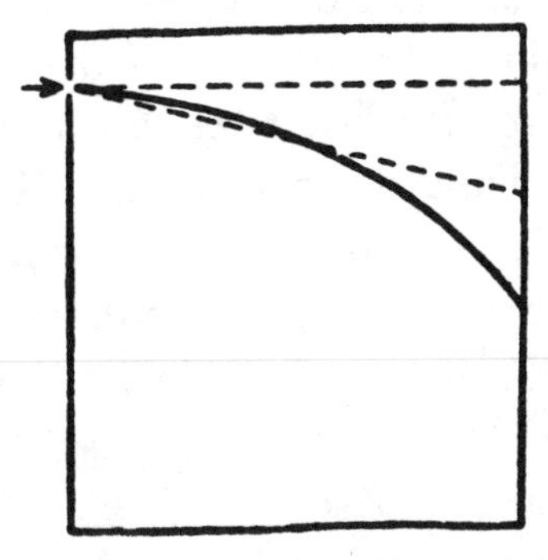

Fig. 9. The Curvature of Light-Rays in Einstein's Box

We shall elucidate this characteristic trend of thought by applying it to a certain example, namely, to the connection of light and gravitation. For this purpose, we turn once more to the box in which the physicist performs his experiments without being able to distinguish between acceleration and gravity.

Let us assume that the box is at rest (Fig. 9). In a side wall there is a small hole through which a ray of light shines in; it follows a straight, horizontal line in the dust of the air (represented by the dotted line of the figure). If the box is now set in uniform motion, the line changes: whereas the light entering through the hole reached previously exactly the opposite point on the wall, now that the box moves up the point of illumination goes further down, away from the ceiling. The ray is seen now as a sloping line, though still running straight. Next, let us imagine that the box moves upward with acceleration. The farther down sinks the ray, the faster goes up the box, so that the ray takes the distorted form of a curved line (see the solid line). In the dust of the air, it would be seen in the shape of a water jet spurting sidewise from the pipe and flowing down in an arc. This experiment cannot, of course, be actually performed, for the simple reason that light propagates so fast that, in contrast to it, the spatial displacement of the box in the same period of time amounts practically to nothing; no change in the ray could be actually observed. Our experiment is supposed to be merely "mental", intended to clarify the principle.

Let us now turn to Einstein's principle of equivalence. Einstein maintains it is immaterial whether we consider an accelerated motion or a gravitational field. It follows: As the curvature of the light rays occurs in the case of accelerated motion, so it must occur also in a gravitational field. The surprising conclusion results immediately from the principle.

We are facing here an entirely new consequence of

Einstein's theory of gravitation. The assertion is of a far-reaching significance. According to it, light does not propagate in open space in a straight line when it comes within the sphere of the attraction of masses; on the contrary, it follows a curved path not unlike that of a flying missile. This contention could be examined astronomically in repeated observations since Einstein deduced it for the first time from his theoretical considerations; and it has been confirmed to its full extent. Such observations not only require great precision but they can be made only during a total eclipse of the sun; elaborate preparations are therefore demanded of the astronomer who wishes to check Einstein's effect.

Einstein has drawn still another conclusion from his principle of equivalence, which concerns the behavior of clocks within the field of gravitation. By calculating certain deviations of the clock for the accelerated motion of the above mentioned box and by transferring the results to gravitational fields, he concluded, on the basis of considerations similar to those just outlined, that a clock, subjected to the influence of a strong gravitational field, would become slow. This effect cannot be demonstrated, of course, on ordinary clocks, as all watches and even the finest chronometers are still too inexact to be used for measuring these small retardations. But the physicist knows another kind of watches the precision of which transcends by far anything of human making: they are the individual atoms of which all substance is constructed. Let us describe briefly the plan for the demonstration of Einstein's doctrine, based on this effect.

Since the investigations of the last decades, it has become known that the atom is not a uniform body, but consists of two distinct kinds of material, the positively charged nucleus and the negatively charged electrones; the heavy but very small nucleus stands in the middle, while electrons revolve round it in their elliptical course. On account of this circular movement of the electrons, the whole atom can be conceived as a clock, in which each revolution of an atom corresponds to one turn of the hand and constitutes a unit of clock-time. Now, the revolution of electrons can be measured very exactly, insofar as it manifests itself in the number of vibrations of the light emitted by a circulating electron. Almost everybody has occasionally observed how a gas-flame becomes colored once salt gets into it; ordinary cooking salt colors the flame yellow, because it contains sodium; potassium colors the flame violet, etc. This coloration is due to the fact that the atoms of basic elements are "stimulated" by the flame and emit light the vibrations of which depend on the number of electronic revolutions, manifesting themselves in the color of the light. The exact estimation of the color is done by means of so-called spectral lines which are observed and photographed in an extremely delicate apparatus, the spectrometer. This apparatus splits every light into its component parts, so that white light is transformed by it into a "spectrum" resembling the color sequence of the rainbow and extending from red to orange, yellow, green, blue, and violet. The lights of the radiating atoms, on the contrary, are marked in fine but sharp transverse lines, separated from each other, and each appearing in one definite color.

Einstein maintains that such an atomic clock manifests retardation in a gravitational field. A very strong gravitational field, a much stronger one than anywhere on the earth, exists on the sun, for the mass of the sun is by far greater than that of the earth. The atmosphere of the sun consists of incandescent gases; as the conditions prevailing there resemble those within the gaseous flame, atoms are aglow. In fact, with the help of a spectral apparatus, it is possible to recognize, as spectral lines, the colors emitted by individual elements of the sun and to measure the number of their vibrations. If the individual atoms are really somewhat retarded in their motion by the gravitational field of the sun, then the spectral lines arising in them must occupy a slightly different position in the spectrum than the lines arising in the earthly sources of light. They must shift in the direction of the lower number of vibrations, that is, toward the red end of the spectrum. One speaks, therefore, of the red shift of the spectral lines, observed in the sunlight.

The experimental test has encountered great difficulties at first, insofar as it deals with an extremely small deviation and the calculated effect lies just on the borderline of the measurable. But recently, very precise measurements have satisfactorily confirmed Einstein's findings. The astronomer, E. Freundlich, in order to

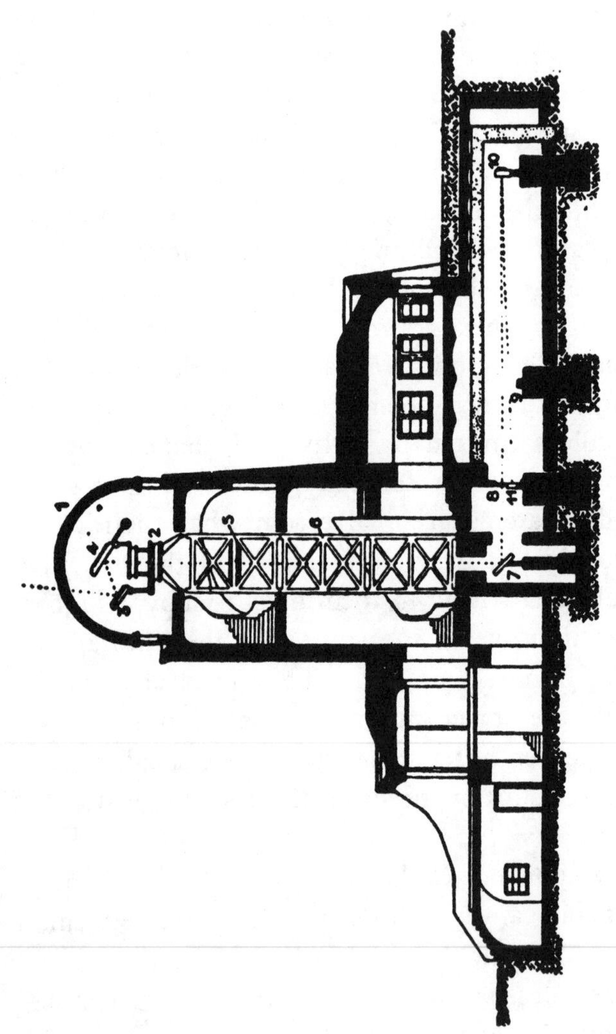

Fig. 10. *The Einstein Tower in Potsdam*

1. Cupola. 2. Revolving style for the mirror. 3. Coelostat. 4. Counter-mirror. 5. Objective. 6. Wooden scaffold. 7. Steering mirror. 8. Slot. 9. Prism apparatus. 10. Diffraction grating. 11. Photographic camera.

reach a conclusive demonstration of this, has built in Potsdam the Einstein tower (shown in Fig. 10), a structure combining to perfection every astronomical and physical contrivance. The tower has a lens (5) in its cupola, into which the light of the sun is directed from a side (mirror system, 3, 4), so that the tower as a whole forms a single large telescope. At the foot of the tower the light is caught (7) and directed toward a huge spectral apparatus. A space several meters long (8-10), which is completely shut off from the surrounding world, forms the interior of the apparatus. At 8 the light enters through a slot; and at 10 is found the most valuable instrument of the whole arrangement, the diffraction grating, consisting of a slightly curved metallic mirror with innumerable and extraordinarily fine scratch-lines. It splits light into its constituent colors and reflects it back to 11, where it is reproduced on photographic plates. The final measurements of the red shift are supposed to begin soon.*

Finally, we wish to mention, in this connection, the third astronomical test found by Einstein for his theory. With the mathematical elaboration of the theory, it became clear that the planetary movements followed a much more complex law than taught by Newton and

* The experiments in the Einstein tower could not be continued since Professor Freundlich was forced to leave Germany when the Hitler government came into power. The Einstein tower was given a new name and is now used for purposes which the Nazi government deems less dangerous for the German race. Up to the present time a definitive clarification of the red shift of spectral lines in the sun has not been given. (*Translator's note.*)

believed since his days. Newton's doctrine, to the effect that the sun attracts the planets with a power decreasing in proportion to the square of the distance, was shown by Einstein to be only approximately correct. It must be replaced, for more exact purposes, by a different law. Whereas every planet, according to Newton, describes an ellipse around the sun, it follows from Einstein's law that, though this ellipse is indeed described, nevertheless it is accompanied by another rotary movement: the ellipse, as a whole, revolves around the sun in the course of centuries. This rotary movement must be strongest for the planets in the neighborhood of the sun. The astronomers had noticed since the middle of the last century, that the planet Mercury shows certain deviations from its course: its ellipse actually executes a rotary movement of the kind. This was found in the lateral retrocession of one of the extreme orbital points, the perihelion. This so-called perihelion movement of Mercury amounts to only 43 seconds of the arc per century. Yet the astronomers were unable to find a satisfactory explanation of the fact. Einstein's law gave an explanation of this rotation of the ellipse.

The coincidence of theory and observation has, in this case, remarkable force of persuasion. It would not be surprising, if a theory devised originally for the explanation of the perihelion movement were to determine correctly the amount of this deviation. However, Einstein's theory has arisen from entirely different grounds. It is based on ideas concerning the relativity of motion, the equivalence of gravity and acceleration; and all its constructions are made in the pursuit of this program. It was, therefore, highly surprising that Einstein, after being informed at a rather late stage of his ideas of the fact of the perihelion movement of Mercury, subjected his theory (rooted in entirely different sources) to the test of whether or not it will give an answer to this question. And when the long known amount of 43 seconds of the arc was deduced from his theory, he had every right to regard this unexpected coincidence as an excellent confirmation of his assumptions.

We have described in the preceding pages the astronomical consequences of the theory of relativity in such detail, because we are interested in showing that facts of observation have been the ultimately deciding factors in the acceptance of Einstein's theory. This is its strength; for, in the last analysis, the final confirmation of physical ideas can be given only by nature itself. Were it merely the question of creating a picture of how to make intelligible the inner workings of nature, physics would be a very simple science. Explanations are found altogether too easily, when imagination is given a little rein. But it is truly an art to find explanations from which new facts follow and which can be confirmed by experiments. This applies, above all, to the numerous inventors who still occupy themselves with the problem of ether and who still look around for ideas as to how to reconcile the contradictory properties of such ether. Such ideas can always be found; but they lack the force of conviction, because their authors do not succeed in getting new experimental results from their theories. It is easy to devise a theory of ether, capable of accounting even for the curvature of light and the red shift; there is no trick to it after these effects have been discovered by Einstein. Whoever believes firmly in the existence of ether should take example from Einstein and *predict* effects capable of experimental proof. But as long as this does not occur and only the prenomena predicted by Einstein are observed, so long shall we adhere to Einstein and to his theory of gravitation, which is also a theory of the relativity of motion.

We do not wish to attempt presenting the mathematical structure of Einstein's theory. Nobody will doubt our words that, mathematically, it is an exceedingly intricate matter. Einstein aimed to find a general concept of gravitation that would fit all the different descriptions which could be given for the state of gravitation. For this purpose, he had to introduce in physics a new mathematical method, the so-called tensor calculus. We are reminded here of Newton's case who, in a similar manner, had to develop a new mathematical method, that of the differential calculus, on which to construct his theory of gravitation. However, whereas Newton had to invent, at that time, the method of calculation himself, Einstein was fortunately able to utilize for this purpose the mathematicians' works which were already available. The essence of the new method of calculation resides in two basic concepts, the *invariant* and the *co-variant.* The field of gravitation is a co-variable magnitude. If one passes from one frame of reference to another, this magnitude changes, varies with—and this is the meaning of the word "co-variant'. Nevertheless, one should not believe that the objective meaning of the knowledge of nature would be eliminated thereby; for all such descriptions given in terms of different frames of reference signify merely different ways of expression, enabling us to comprehend the true character of nature. It is something like the way in which one can express thought in German, English, French, etc.; the language may be different, but the mental content is the same. Similarly, the presentation of the state of gravitation in the world can be made in different languages, depending on the chosen frame of reference. But all these descriptions refer to one and the same objective state. This state is the invariable, the unchangeable. The peculiarity of the mathematics of relativity is perhaps best expressed in this pair of concepts, the invariant and the co-variant. The co-variant stands for the manner of description; the invariant, for the common state arrived at from all the various descriptions.

It is important to make this thought clear. It is occasionally attempted to present Einstein's theory in the simple sentence that everything is relative. But Einstein has not made everything relative. Only some things have become relative, particularly things previously regarded as absolute verities. On the other hand, the theory has made only clearer the things which are true regardless

Chapter 6 : SPACE AND TIME

IN THE preceding chapters we have described the physical side of the discoveries connected with the theory of relativity. In doing so, we put a special emphasis on factual foundations, that is, on the data of observation and experimentation, which gave rise to the bold conclusions drawn by Einstein. In this last chapter, we intend to consider the other side of the problem, dealing not so much with physics as with another realm, that of philosophy. Our theory will appear, in this light, no less important and significant. We encounter here the thoughts which made the theory of relativity famous in wide circles, which distinguish it from other physical theories and secure for it a prominent position within the modern philosophy of nature. It is the revolution of our ideas concerning space and time, to which we turn with this analysis.

As far as time is concerned, a substantial part of the new ideas has already been presented in the chapter on the special theory of relativity. The foremost place is occupied here by the relativity of simultaneity; it maintains that the time-order of events separated by distance is arbitrary within certain limits. It must be stressed once more that the events in question must be widely separated in space. We have found that the time-order of such events is not accessible to direct observation. As observers, we can be in the neighborhood only of one of the events; a signal must be sent from the other event, which thus notifies us of the event's existence. If we wish to be informed as to the time at which it occurred, we must resort to calculation; for that we must know the velocity of the signal. Yet we have found that it is impossible to measure the velocity, unless we have already established simultaneity; for such a measurement requires two clocks, correctly set and placed at different localities. The argument thus runs in a circle, one premise presupposing the other; and its solution consists in abandoning the objective meaning of simultaneity. Simultaneity cannot be known, it must be defined, and this definition will be arbitrary to a certain extent. If cannons were fired on two distant mountains at the same time, I should hear the two reports simultaneously only if I were standing in the middle of the distance. I then could assert also that the two discharges did not occur simultaneously but in succession; and that could be justified by ascribing to sound waves a greater speed in one direction than in the other. I could then consider, quite arbitrarily, one or the other discharge as the earlier. Such an assertion would never involve me in contradiction; for I shall always be able to account for my observation: namely, that I hear the two reports simultaneously in the middle of the distance.

Here lies one of the deepest thoughts of the theory of relativity. We shall regard as true whatever we observe immediately; no theory can put out of existence whatever our senses teach us. An unconditional respect for the evidence of the senses, of experience, constitutes the basic principle of the theory of relativity. This is supplemented, however, by the clear realization that the power of human observation is limited. Only a small portion of the world-space can be mastered by the senses; whatever happens beyond it, must be deduced by reflection. This is where reasoning comes in; by its force our knowledge expands beyond the narrow horizon of vision and opens up before us the gates of distant worlds. When we declare that we see the stars, this is a very inexact way of expression; we see directly only the light penetrating our eye. If we proceed from the experience of brightness, occurring here, to the statement that there are stars far away, we are compelled to draw an inference; and this inference cannot be drawn without some arbitrariness. One part of this arbitrariness is represented by simultaneity. The way we define it can change our system of thought, but it cannot change the observed facts themselves; that is why all these different descriptions are equally true and equally justified.

The relativity of simultaneity has a peculiar consequence, as far as the measurement of space is concerned. We shall make this clear by means of an instructive example. For this purpose we consider an apparatus, well-known in photographic practice, the so-called focal-plane shutter.

Most photographic cameras are equipped with a shutter mounted between the lenses; but all these shutters prove to be inadequate for the photography of fast moving objects, because their exposure time cannot be made short enough. A focal plane shutter is used, therefore, for very short exposures. In such a camera there runs vertically outward, close to the film, and therefore practically in the focal plane, a rolling curtain with a horizontal slit in it; the various parts of the film receive light only as long as the slit passes them. The time of exposure is, therefore, extremely short. But at the same time a peculiar fault creeps in: the individual sections of the plate do not receive light all at the same time, but only one after another, and as the object moves while being photographed the individually illuminated sections do not

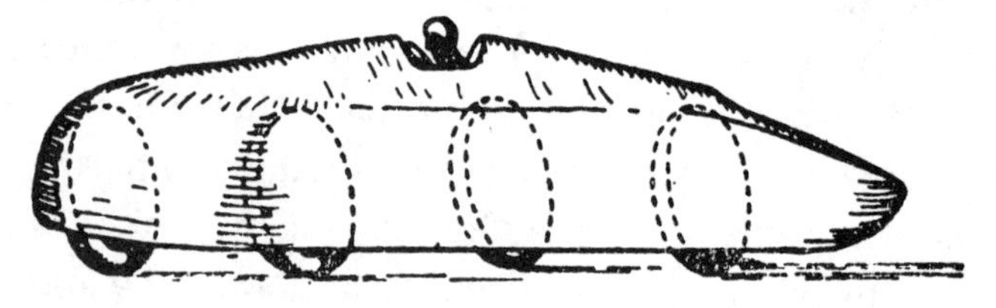

Fig. 11. *Major Segrave's* 1,000 *Horsepower Auto at Full Speed*

represent strictly simultaneous states of the object, but successive states. The object cannot change very much, however, in that brief period of time; nevertheless, a certain distortion of the picture does occur. This can be well observed on the wheels of a fast moving automobile, since they assume the shape of a somewhat crooked ellipse with a forward tilt (Fig. 11).

A similar distortion occurs, according to Einstein, when one wants to determine the shape of moving bodies.

The difficulties found here were not seen at all before Einstein. For if one observes a moving body from a frame of reference at rest, the moving object is "photographed", so to speak, from a position at rest; and then the image is examined. The moving body appears to an observer at rest as a sequence of such instantaneous snap-shots. At this point the relativity of simultaneity comes into consideration; events which are conceived as simultaneous for one definition of simultaneity, represent a sequence of time for another. The significance of this, as far as pictures of moving bodies are concerned, is as follows: what is instataneous photography for one temporal system, is a photography by focal plane shutter for another. The shape of moving bodies varies according to the definition of simultaneity. There are no true shapes of moving bodies; all shapes obtainable in this way are equally true.

This is Einstein's theory of the change in the form of moving bodies. The comparison with a photography by focal plane shutter represents the nature of this theory extremely well. The only difference consists in that Einstein's focal plane shutter would have to run faster than light. It therefore cannot be actualized by such an apparatus as a photographic shutter. On the other hand, it follows from this fact that Einstein's "distorted snapshots" are not "false"; they can just as well be considered as strictly instantaneous snapshots. This result does not hold for ordinary photography by focal plane shutter; pictures so obtained must rightly be called distorted.

Our reflection shows us that space-measurement depends on simultaneity. This idea can be expressed mathematically by bringing together space and time into a four-dimensional structure, into a space-time manifoldness. Strangely enough, this procedure which appears simple and harmless to the mathematician, has given cause for great surprise and for bewilderment to others. Many a reader of books on relativity thought that space was thereby transformed from a three-dimensional structure into a four-dimensional one; and he then attempted in vain to conceive the fourth dimension of space. He may have argued in this way: Imagine three sticks of wood meeting together at one point under right angles, like the length, width and height of a room. These are three dimensions of space; is there any room for the fourth one? How is it possible to pass the fourth stick through the point, so that it too would form right angles with the others? The author too cannot visualize how it would run; but the theory of relativity never asserted anything of the sort. It asserts merely that time should be added, *as time*, to space; and this is something entirely different. We may imagine it this way: Three numbers are needed to determine a point in space. Suppose a lamp hangs in the room. How can we determine its place? We measure its distance from the floor, from the back-wall and from the side-wall; these three figures determine its position in space. The three numbers are called co-ordinates. The room is three-dimensional, because three figures are needed for statements of the kind described. If we want to determine not a point in space but an event, we require another figure, namely, the statement of time. Suppose that we switch on the light for a second and produce a flash of light; this is an event. It is completely determined if we know the three numbers defining the position of the lamp and, in addition, the fourth number defining the time of the light-flash. Insofar as there are four figures, space and time together are called a four-dimensional manifoldness. This is the whole secret. Unfortunately, this simple circumstance is often depicted in a most obscure language.

Whatever new is asserted by the theory of relativity about the space-time manifoldness, is illustrated much more comprehensibly and clearly in our picture of the focal plane shutter. It shows that the measurement of space is dependent on the measurement of time. This is, of course, something very new and profound; but it does not deprive time of its specific temporal character. Rather, it must be said that only the theory of relativity has discovered and formulated the peculiar distinction of time and space. The philosophical investigation of the theory of relativity has shown that time is something even more profound than space, that it is connected with the deepest principle of all knowledge of nature, the law of cause and effect.

If we now turn to the problem of space, we find here ideas going farther back than the relativistic doctrine of time. For what Einstein teaches about space and geometry, has been prepared, on the mathematical side, one hundred years ago. These ideas are connected with the so-called non-Euclidian geometry. The geometry studied by us in school goes back to the Greek mathematician, Euclid; it has been taught for two thousand years in the form originally given by him. Only within the last century a new kind of geometry was discovered by several mathematicians, among whom Riemann is the most important. This geometry appears at first glance totally unreasonable and nonsensical, insofar as it contains such sentences as that the three angles of a triangle are together more than 180°, or that the circumference and diameter of a circle do not stand in the relationship $\pi = 3.14$. A more exact examination, however, proves it to be a completely correct and permissible mathematical system, to which one has only to get used.

The non-Euclidian geometry may be conceived simply as a play with concepts which, though logical in themselves, have no significance beyond that. It seemed in fact that real space, the space of things and bodies of the universe, followed the laws of old Euclidian geometry. These laws were always taken as basic, whenever houses and streets were built, or areas measured for topographic maps, or cosmic distances calculated. But already the discoverers of non-Euclidian geometry asked themselves the question as to whether Euclid's laws are strictly true; possibly, they thought, more exact measurements may bring to light deviations corresponding to non-Euclidian geometry. They knew full well that such deviations can be expected only for very large dimensions. The great mathematician, Gauss, undertook therefore to measure a triangle of large size. The corner-points of his triangle were formed by three mountains: Brocken in Harz, Inselsberg in the Thueringian forest, and Hohenhagen near Goettingen. The summits of these mountains were almost at the limit of visibility from each other, if telescopes were used. Gauss measured the three angles enclosed by this triangle and inquired whether their sum differed from 180°; however, there was no noticeable deviation. Nevertheless, some mathematicians and physicists believed ever since then that some day a deviation may be revealed in still larger triangles by means of more precise instruments.

The relations governing space, in that case, can be elucidated if we take as our starting point the corresponding relations in two-dimensional surfaces. It is found that the laws similar to those holding for non-Euclidian geometry of three-dimensional space actually apply to such two-dimensional structures as curved surfaces. At the same time, let us depict much greater deviations than those assumed in Gauss's experiment; it then will be easier to visualize the relations to be considered.

Let us imagine beings living on the surface of a globe, for whom nothing exists outside this globe-surface. In their world, there would not be any tunnel going through the globe; nor would it include things stretching away from the globe, such as trees or towers. Everything is flat for them, embedded completely in the surface of the sphere, including the beings tehemselves. Now the question arises: would these beings be capable of noticing that they live on a curved surface?

The answer to this question is by no means self-evident. We notice the curvature of the surface of the earth mainly because we observe phenomena outside the two-dimensional surface. When we observe the curvature of a hollow in the ground we *sight* across it, i.e., we compare its form with the course of light-rays; we see the curvature of the hollow merely because light is not confined to the curved surface but freely permeates the three-dimensional space. But in the two-dimensional world as conjectured, light-rays would glide along the surface; therefore no curvature would be noticed by sighting. And yet there would be other ways to recognize the curvature.

Suppose that those living beings undertake surveying; they draw figures in the sand and measure them with yardsticks. They draw a circle around the north pole of the globe, for instance, a circle corresponding to 89° of northern latitude. Then they measure the circumference of the circle, using the yardstick. Finally, they measure the diameter of the circle; but what will they measure as diameter? Certainly not the "true" diameter traversing the interior of the sphere, along the chord; for they cannot leave the surface of the globe, and there does not exist anything for them outside the surface. Consequently, they will take for diameter the curved line running from one point of the circle by the north pole to its opposite point. This line will appear straight to them, because, in following it with the eye, they see the opposite point, insofar as light moves along the contour of the globe. But, if they measure the length of this line by using the yardstick, and then divide the circumference of the circle by the figure obtained for the diameter, they will get a smaller number than $\pi = 3.14$, as the measure of the diameter is too large. By the results of these measurements they will know that they live on the surface of a globe.

Now let us describe the corresponding situation for three dimensions. Suppose there is a large sphere of iron sheet, about the size of a house. There is an iron scaffold inside. A man climbs on it; he can climb also the outer surface, where there are handles and steps to cling to. He measures the circumference of the sphere with a yardstick and then the diameter in a similar way, climbing along one of the girders. Finally, he divides the figures and gets a smaller number than $\pi = 3.14$.

The result was easy to understand in the case of two dimensions. The surface was conceived as curved or bent in the third dimension, as a sphere's surface must be. But for the case of three dimensions, this answer is no longer possible. There is no room for curving the three-dimensional space. How shall we then interpret the result? Nothing remains for us to do but to admit that we live in a non-Euclidian space. Those experiences in measuring are what would be noticed in such a space as space-curvature. Furthermore, we must keep in mind that the described two-dimensional creatures would have no other way of visualizing the curvature of their two-dimensional space; they cannot speak of its bending in the third dimension. The deviation from normal measuring conditions is just what one would experience inside a non-Euclidian space.

We cannot go here any further into the problem of visualizing non-Euclidian space; for a more detailed treatment of these questions, we must refer the reader to the author's *Philosophy of Space and Time*,* which in general must be consulted for a more extensive explanation of the thoughts contained in this book. There we discuss, in particular, the question of the relativity of geometry; it appears, namely, that all geometrical measurements imply an uncertainty similar to that of the relativity of motion, and that measurements of the objective geometry of space presuppose a special sort of definitions which we call coordinative definitions. This question is connected with the question of whether there exists a Euclidian interpretation of measurements as described. Here we must face the question as to how Einstein came to apply non-Euclidian geometry to his theory of gravitation.

We have already pointed out in Chapter 3 that watches and yardsticks have no independent significance, according to Einstein's conception, but change in a particular way and are adjusted to the geometry of light. But even light is not the final thing; for it, too, is subjected to the guiding power of gravitation. It may be well to remind

*H. Reichenbach, *The Philosophy of Space and Time,* English translation, Maria Reichenbach and John Freund, Dover Publications, Inc., New York, 1957. Cf. also H. Reichenbach and E. S. Allen, *Atom and Cosmos: The World of Modern Physics,* Ridgeway Books, Philadelphia, 1933.

here of the argument contained in Chapter 5, according to which light conforms to the gravitational field. Gravitation is the primary effect of the masses filling space; it is the guiding power to which light, yardsticks and watches conform. The simple relations of spatial measurement, as formulated in Euclidian geometry, are valid only in the absence of a gravitational field, that is, at great distances from the star masses. In the vicinity of such great masses, on the other hand, space is warped, so to speak; it assumes curved forms and follows strange laws, as given in non-Euclidian geometry. The deviation from Euclidian relations is always, to be sure, very small, so small, in fact, that it cannot be demonstrated by means of ordinary measuring devices. This is the reason why it passed so long unnoticed. Even such measurements as those of Gauss could lead to no success, because they invariably dealt with too small distances. The deviations manifest themselves only in cosmic distances; and it is the course of heavenly bodies and of light-rays between them that betrays the non-Euclidian nature of space. And there, in the wide stretches of the universe, we find, indeed, quite substantial changes of geometry.

The most perplexing thing of it all is that the space of the universe must now be considered as finite. This does not mean that the masses of the stars alone are finite; it means that space itself is limited. We can visualize this in the following manner. If a ray of light is sent out in a straight line, it returns after a certain time from the opposite side, not unlike a ship sailing steadily west but returning to the port of departure from the other side. There is no unlimited extension in this space; all straight lines come finally to their source. Each star can be potentially seen twice, therefore, once from the front and the second time from behind, when we look at it about the universe. Unfortunately, no proof of this theory of Einstein can be given at the moment, for the road around the world is so long that the stars' light grows too weak to be observed. But even if we could see the light, there would be no way of recognizing the particular star. In the countless thousands of years required by light to go around the world, the star would have wandered far away and would occupy an entirely different position from its counterpart; as a result, we should not be able to recognize the two stars as identical.

Einstein's conception of gravitation as a "metric power", as a force determining the relations of spatial measurement, leads therefore to a far-reaching revolution in our knowledge of space. Apart from the novelty of the theory of a limited heavenly space, which signifies a turning point similar to that of the doctrine of the spherical shape of the earth, at the time of its promulgation, the method of dealing with the problem of space, applied in Einstein's theory, represents a new form of philosophical thinking. It follows the principle that statements concerning space are not to be separated from statements concerning bodies in space, that a space has no absolute significance apart from things and the laws of their mutual relations, a principle recognized before

Einstein only by Leibniz. This limitation of the concept of space to its bodily manifestations represents a key to the understanding of the meaning of geometry, a problem which, after the discovery of non-Euclidian geometry, could no longer be solved by Kant's doctrine of an apriori validity of Euclidian geometry. The apparent priority of the latter geometry, expressed in the fact that it controls all our spatial imagery, can be understood if we realize that the space-perception we possess has arisen historically from contact with things following the laws of Euclidian space. The solid bodies and sticks we work with comply so closely with the rules of Euclidian geometry that we do not notice any deviations from it; as a result, we have become so accustomed to the laws of Euclid that we regard them as absolutely necessary. The deviations pointed out by Einstein occur only in astronomic dimensions. Were we to live, however, in a world where the laws just described should hold in the dimensions of our daily environment — where, for example, the measured relations between circumference and diameter would differ from 3.14 — we should get accustomed also to these facts. We should find everything self-evident and natural. If a physicist came along and asserted the opposite, namely, that Euclidian geometry must determine all our spatial imaginations, we should answer him that he asserts the impossible; and his loudest opponents would be the very persons who defend today the apriori character of Euclidian geometry. The great achievement of Einstein consists in that his thinking is free from conventional ideas, that he did not hesitate to disregard the oldest laws of natural science, the laws of geometry, and to set new ones in their place. Though these new geometrical laws were recognized by other mathematicians before him, Einstein was the first one to take them down from the shelves of thought-possibilities and to apply them to physical science, to the description of nature. Such a scientific deed manifests boldness, reveals independence of thought; and we should not be astonished that it was difficult for all of us, and will be so for every one who hears of these ideas for the first time, to understand Einstein's theory.

Once more a chapter of our presentation ends with a Copernican turn. The first such turn was given by the demonstration of the relativity of motion; with this principle the step from the Ptolemaic world view to the Copernican one was repeated on a higher level, leading to a synthesis of both world views into one. In a similar way, the break with Euclidian geometry shakes the very foundations of our knowledge and signifies a transition to a knowledge of a higher kind, incomprehensible as this knowledge may appear at first view. But just as the Copernican worldview became at last generally recognized and a common property of all educated people, so will it be with the theory of relativity. One hundred years from now, the doctrine will be accepted as self-evident; and it will be difficult to comprehend why it encountered at first so much opposition. In Schopenhauer's words, "Truth is allowed only a brief interval of victory between the two long periods when it is condemned as paradox or belittled as trivial." We who are permitted to see this period of victory with our own eyes may consider ourselves fortunate to witness the Copernican discovery of our age.

A CATALOGUE OF SELECTED DOVER BOOKS
IN ALL FIELDS OF INTEREST

A CATALOG OF SELECTED DOVER BOOKS IN ALL FIELDS OF INTEREST

LASERS AND HOLOGRAPHY, Winston E. Kock. Sound introduction to burgeoning field, expanded (1981) for second edition. 84 illustrations. 160pp. 5⅜ × 8¼. (EUK) 24041-X Pa. $3.50

FLORAL STAINED GLASS PATTERN BOOK, Ed Sibbett, Jr. 96 exquisite floral patterns—irises, poppie, lilies, tulips, geometrics, abstracts, etc.—adaptable to innumerable stained glass projects. 64pp. 8¼ × 11. 24259-5 Pa. $3.50

THE HISTORY OF THE LEWIS AND CLARK EXPEDITION, Meriwether Lewis and William Clark. Edited by Eliott Coues. Great classic edition of Lewis and Clark's day-by-day journals. Complete 1893 edition, edited by Eliott Coues from Biddle's authorized 1814 history. 1508pp. 5⅜ × 8½.
21268-8, 21269-6, 21270-X Pa. Three-vol. set $22.50

ORLEY FARM, Anthony Trollope. Three-dimensional tale of great criminal case. Original Millais illustrations illuminate marvelous panorama of Victorian society. Plot was author's favorite. 736pp. 5⅜ × 8½. 24181-5 Pa. $8.95

THE CLAVERINGS, Anthony Trollope. Major novel, chronicling aspects of British Victorian society, personalities. 16 plates by M. Edwards; first reprint of full text. 412pp. 5⅜ × 8½. 23464-9 Pa. $6.00

EINSTEIN'S THEORY OF RELATIVITY, Max Born. Finest semi-technical account; much explanation of ideas and math not readily available elsewhere on this level. 376pp. 5⅜ × 8½. 60769-0 Pa. $5.00

COMPUTABILITY AND UNSOLVABILITY, Martin Davis. Classic graduate-level introduction th theory of computability, usually referred to as theory of recurrent functions. New preface and appendix. 288pp. 5⅜ × 8½. 61471-9 Pa. $6.50

THE GODS OF THE EGYPTIANS, E.A. Wallis Budge. Never excelled for richness, fullness: all gods, goddesses, demons, mythical figures of Ancient Egypt; their legends, rites, incarnations, etc. Over 225 illustrations, plus 6 color plates. 988pp. 6⅛ × 9¼. (EBE) 22055-9, 22056-7 Pa., Two-vol. set $20.00

THE I CHING (THE BOOK OF CHANGES), translated by James Legge. Most penetrating divination manual ever prepared. Indispensable to study of early Oriental civilizations, to modern inquiring reader. 448pp. 5⅜ × 8½.
21062-6 Pa. $6.50

THE CRAFTSMAN'S HANDBOOK, Cennino Cennini. 15th-century handbook, school of Giotto, explains applying gold, silver leaf; gesso; fresco painting, grinding pigments, etc. 142pp. 6⅛ × 9¼. 20054-X Pa. $3.50

AN ATLAS OF ANATOMY FOR ARTISTS, Fritz Schider. Finest text, working book. Full text, plus anatomical illustrations; plates by great artists showing anatomy. 593 illustrations. 192pp. 7⅛ × 10¼. 20241-0 Pa. $6.00

EASY-TO-MAKE STAINED GLASS LIGHTCATCHERS, Ed Sibbett, Jr. 67 designs for most enjoyable ornaments: fruits, birds, teddy bears, trumpet, etc. Full size templates. 64pp. 8¼ × 11. 24081-9 Pa. $3.95

TRIAD OPTICAL ILLUSIONS AND HOW TO DESIGN THEM, Harry Turner. Triad explained in 32 pages of text, with 32 pages of Escher-like patterns on coloring stock. 92 figures. 32 plates. 64pp. 8¼ × 11. 23549-1 Pa. $2.50

CATALOG OF DOVER BOOKS

THE PRINCIPLE OF RELATIVITY, Albert Einstein et al. Eleven most important original papers on special and general theories. Seven by Einstein, two by Lorentz, one each by Minkowski and Weyl. 216pp. 5⅜ × 8½. 60081-5 Pa. $3.50

PINEAPPLE CROCHET DESIGNS, edited by Rita Weiss. The most popular crochet design. Choose from doilies, luncheon sets, bedspreads, apron—34 in all. 32 photographs. 48pp. 8¼ × 11. 23939-X Pa. $2.00

REPEATS AND BORDERS IRON-ON TRANSFER PATTERNS, edited by Rita Weiss. Lovely florals, geometrics, fruits, animals, Art Nouveau, Art Deco and more. 48pp. 8¼ × 11. 23428-2 Pa. $1.95

SCIENCE-FICTION AND HORROR MOVIE POSTERS IN FULL COLOR, edited by Alan Adler. Large, full-color posters for 46 films including *King Kong, Godzilla, The Illustrated Man,* and more. A bug-eyed bonanza of scantily clad women, monsters and assorted other creatures. 48pp. 10¼ × 14¼. 23452-5 Pa. $8.95

TECHNICAL MANUAL AND DICTIONARY OF CLASSICAL BALLET, Gail Grant. Defines, explains, comments on steps, movements, poses and concepts. 15-page pictorial section. Basic book for student, viewer. 127pp. 5⅜ × 8½. 21843-0 Pa. $2.95

STORYBOOK MAZES, Dave Phillips. 23 stories and mazes on two-page spreads: *Wizard of Oz, Treasure Island, Robin Hood,* etc. Solutions. 64pp. 8¼ × 11. 23628-5 Pa. $2.25

PUNCH-OUT PUZZLE KIT, K. Fulves. Engaging, self-contained space age entertainments. Ready-to-use pieces, diagrams, detailed solutions. Challenge a robot, split the atom, more. 40pp. 8¼ × 11. 24307-9 Pa. $3.50

THE HUMAN FIGURE IN MOTION, Eadweard Muybridge. Over 4500 19th-century photos showing stopped-action sequences of undraped men, women, children jumping, running, sitting, other actions. Monumental collection. 390pp. 7⅞ × 10⅝. 20204-6 Clothbd. $18.95

PHOTOGRAPHIC SKETCHBOOK OF THE CIVIL WAR, Alexander Gardner. Reproduction of 1866 volume with 100 on-the-field photographs: Manassas, Lincoln on battlefield, slave pens, etc. 224pp. 10⅜ × 8¼. 22731-6 Pa. $6.95

FLORAL IRON-ON TRANSFER PATTERNS, edited by Rita Weiss. 55 floral designs, large and small, realistic, stylized; poppies, iris, roses, etc. Victorian, modern. Instructions. 48pp. 8¼ × 11. 23248-4 Pa. $1.95

AUTOBIOGRAPHY: The Story of My Experiments with Truth, Mohandas K. Gandhi. Boyhood, legal studies, purification, the growth of the Satyagraha (nonviolent protest) movement. Critical, inspiring work of the man who freed India. 480pp. 5⅜ × 8½. 24593-4 Pa. $6.95

ON THE IMPROVEMENT OF THE UNDERSTANDING, Benedict Spinoza. Also contains *Ethics, Correspondence,* all in excellent R Elwes translation. Basic works on entry to philosophy, pantheism, exchange of ideas with great contemporaries. 420pp. 5⅜ × 8½. 20250-X Pa. $5.95

Prices subject to change without notice.

Available at your book dealer or write for free catalog to Dept. GI, Dover Publications, Inc., 31 East 2nd St. Mineola, N.Y. 11501. Dover publishes more than 175 books each year on science, elementary and advanced mathematics, biology, music, art, literary history, social sciences and other areas.